Earth's Biosphere

Earth's Biosphere

Evan Blair

Larsen & Keller
www.larsen-keller.com

Earth's Biosphere
Evan Blair
ISBN: 978-1-64172-669-6 (Hardback)

Larsen & Keller
Published by Larsen and Keller Education,
5 Penn Plaza,
19th Floor,
New York, NY 10001, USA

Cataloging-in-Publication Data

Earth's biosphere / Evan Blair.
p. cm.
Includes bibliographical references and index.
ISBN 978-1-64172-669-6
1. Biosphere. 2. Earth (Planet). 3. Geobiology. 4. Life (Biology). I. Blair, Evan.
QH343.4 .E27 2022
577--dc23

For more information regarding Larsen and Keller Education and its products, please visit the publisher's website www.larsen-keller.com

Table of Contents

Preface

This book aims to help a broader range of students by exploring a wide variety of significant topics related to this discipline. It will help students in achieving a higher level of understanding of the subject and excel in their respective fields. This book would not have been possible without the unwavered support of my senior professors who took out the time to provide me feedback and help me with the process. I would also like to thank my family for their patience and support.

The biosphere is a closed system which includes all the other ecosystems. It is a self-regulating system. It encompasses all living beings and their interaction with the elements of the lithosphere, hydrosphere, geosphere and atmosphere. The main characteristics of the biosphere are the continuous cycle of matter and the flow of solar energy. All of the living organisms of the biosphere depend on water. The biotic components of the biosphere are broadly classified into three categories, namely, primary producers, consumers and decomposers. The topics included in this book on biosphere are of utmost significance and bound to provide incredible insights to readers. Such selected concepts that redefine this discipline have been presented in it. This book is an essential guide for both academicians and those who wish to pursue this discipline further.

A brief overview of the book contents is provided below:

Chapter - Earth's Spheres

The area near the surface of the Earth can be divided into separate interconnected spheres. These spheres are called lithosphere, hydrosphere, biosphere and atmosphere. This is an introductory chapter that discusses the significant aspects of these spheres of the Earth.

Chapter - Biosphere of the Earth

Biosphere can be considered to be the worldwide sum of all ecosystems, which incorporates all the living beings and their relationships with each other. The diverse aspects related to the Earth's biosphere such as the significance of water in it have been thoroughly discussed in this chapter.

Chapter - Biomes

Biomes are the major communities, which are classified on the basis of predominant vegetation and characterized by the adaptation of organisms to that specific environment. Some of the different types of biomes are marine biomes, human micro biomes, endolithic biomes, anthropogenic biomes, etc. This chapter discusses in detail these types of biomes.

Chapter - Classification of Biosphere

Biosphere refers to the global ecosystem of living organisms and non-living factors. It can be classified into various zones such as longitudes and latitudes, Equator, deep biosphere, Arctic circles and Antarctic circles. The aim of this chapter is to explore the classifications of biosphere, which are crucial for understanding the subject in its entirety.

Chapter - Flora and Fauna in the Biosphere

Flora refers to the entirety of plants found in a particular region. Fauna is a term used to represent animal life which is indigenous to a specific region. This chapter has been carefully written to provide an extensive understanding of the importance of flora and fauna in the biosphere.

Evan Blair

1

Earth's Spheres

The area near the surface of the Earth can be divided into separate interconnected spheres. These spheres are called lithosphere, hydrosphere, biosphere and atmosphere. This is an introductory chapter that discusses the significant aspects of these spheres of the Earth.

BIOSPHERE

Biosphere is the thin life-supporting stratum of Earth's surface, extending from a few kilometres into the atmosphere to the deep-sea vents of the ocean. The biosphere is a global ecosystem composed of living organisms (biota) and the abiotic (non-living) factors from which they derive energy and nutrients.

Before the coming of life, Earth was a bleak place, a rocky globe with shallow seas and a thin band of gases—largely carbon dioxide, carbon monoxide, molecular nitrogen, hydrogen sulfide, and water vapour. It was a hostile and barren planet. This strictly inorganic state of the Earth is called the geosphere; it consists of the lithosphere (the rock and soil), the hydrosphere (the water), and the atmosphere (the air). Energy from the Sun relentlessly bombarded the surface of the primitive Earth, and in time—millions of years—chemical and physical actions produced the first evidence of life: formless, jellylike blobs that could collect energy from the environment and produce more of their own kind. This generation of life in the thin outer layer of the geosphere established what is called the biosphere, the "zone of life," an energy-diverting skin that uses the matter of the Earth to make living substance.

The biosphere is a system characterized by the continuous cycling of matter and an accompanying flow of solar energy in which certain large molecules and cells are self-reproducing. Water is a major predisposing factor, for all life depends on it. The elements carbon, hydrogen, nitrogen, oxygen, phosphorus, and sulfur, when combined as proteins, lipids, carbohydrates, and nucleic acids, provide the building blocks, the fuel, and the direction for the creation of life. Energy flow is required to maintain the structure of organisms by the formation and splitting of phosphate bonds. Organisms are cellular in nature and always contain some sort of enclosing membrane structure, and all have nucleic acids that store and transmit genetic information.

All life on Earth depends ultimately upon green plants, as well as upon water. Plants utilize sunlight in a process called photosynthesis to produce the food upon which animals feed and to provide, as a by-product, oxygen, which most animals require for respiration. At first, the oceans and the lands were teeming with large numbers of a few kinds of simple single-celled organisms, but slowly plants and animals of increasing complexity evolved. Interrelationships developed so that certain plants grew in association with certain other plants, and animals associated with the plants and with one another to form communities of organisms, including those of forests, grasslands, deserts, dunes, bogs, rivers, and lakes. Living communities and their nonliving environment are inseparably interrelated and constantly interact upon each other. For convenience, any segment of the landscape that includes the biotic and abiotic components is called an ecosystem. A lake is an ecosystem when it is considered in totality as not just water but also nutrients, climate, and all of the life contained within it. A given forest, meadow, or river is likewise an ecosystem. One ecosystem grades into another along zones termed ecotones, where a mixture of plant and animal species from the two ecosystems occurs. A forest considered as an ecosystem is not simply a stand of trees but is a complex of soil, air, and water, of climate and minerals, of bacteria, viruses, fungi, grasses, herbs, and trees, of insects, reptiles, amphibians, birds, and mammals.

Stated another way, the abiotic, or nonliving, portion of each ecosystem in the biosphere includes the flow of energy, nutrients, water, and gases and the concentrations of organic and inorganic substances in the environment. The biotic, or living, portion includes three general categories of organisms based on their methods of acquiring energy: the primary producers, largely green plants; the consumers, which include all the animals; and the decomposers, which include the microorganisms that break down the remains of plants and animals into simpler components for recycling in the biosphere. Aquatic ecosystems are those involving marine environments and freshwater environments on the land. Terrestrial ecosystems are those based on major vegetational types, such as forest, grassland, desert, and tundra. Particular kinds of animals are associated with each such plant province.

Ecosystems may be further subdivided into smaller biotic units called communities. Examples of communities include the organisms in a stand of pine trees, on a coral reef, and in a cave, a valley, a lake, or a stream. The major consideration in the community is the living component, the organisms; the abiotic factors of the environment are excluded.

A community is a collection of species populations. In a stand of pines, there may be many species of insects, of birds, of mammals, each a separate breeding unit but each dependent on the others for its continued existence. A species, furthermore, is composed of individuals, single functioning units identifiable as organisms. Beyond this level, the units of the biosphere are those of the organism: organ systems composed of organs, organs of tissues, tissues of cells, cells of molecules, and molecules of atomic elements and energy. The progression, therefore, proceeding upward from atoms and energy, is toward fewer units, larger and more complex in pattern, at each successive level.

The Diversity of Life

The biosphere supports between 3 and 30 million species of plants, animals, fungi, single-celled prokaryotes such as bacteria, and single-celled eukaryotes such as protozoans. Of this total, only about 1.4 million species have been named so far, and fewer than 1 percent have been studied for their ecological relationships and their role in ecosystems. A little more than half the named species are insects, which dominate terrestrial and freshwater communities worldwide; the laboratories of systematists are filled with insect species yet to be named and described. Hence, the relationships of organisms to their environments and the roles that species play in the biosphere are only beginning to be understood.

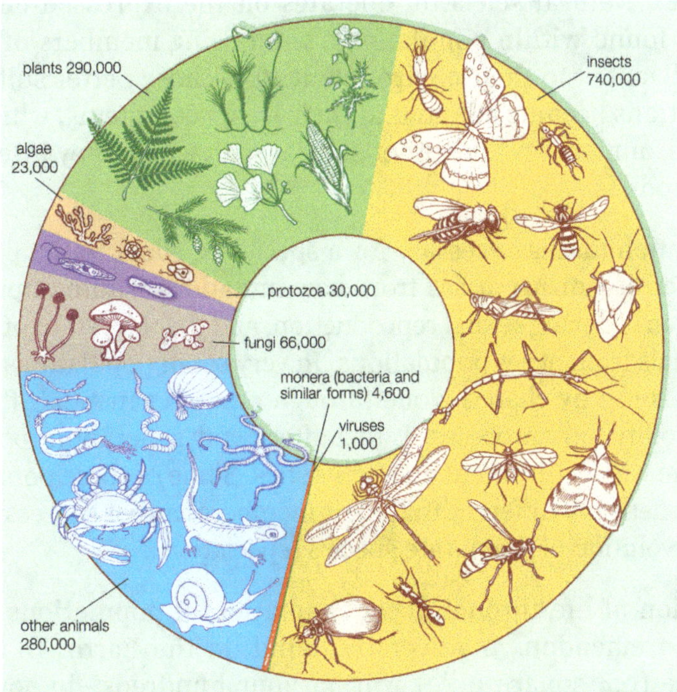

Estimated number of known living species. The majority of species are still unknown—i.e., yet to be described by taxonomists.

The Organization of the Biosphere

Natural Groupings

This tremendous diversity of life is organized into natural ecological groupings. As life has evolved, populations of organisms have become separated into different species that are reproductively isolated from one another. These species are organized through their interrelationships into complex biological communities. The interactions in these communities affect, and are affected by, the physical environments in which they occur, thereby forming ecosystems through which the energy and nutrients necessary for life flow and cycle. The mix of species and physical environments vary across the globe, creating ecological communities, or biomes, such as the boreal forests of North America and Eurasia and the rainforests of the tropics. The sum total of the richness of these biomes is the biosphere.

Processes of Evolution

This hierarchical organization of life has come about through the major processes of evolution—natural selection (the differential success of the reproduction of hereditary variations resulting from the interaction of organisms with their environment), gene flow (the movement of genes among different populations of a species), and random genetic drift (the genetic change that occurs in small populations owing to chance). Natural selection operates on the expressed characteristics of genetic variants found within populations, winnowing members of the population who are less well suited to their environment from those better suited to it. In this manner, populations become adapted to their local ecosystems, which include both the physical environment and the other species with which they interact in order to survive and reproduce.

The genetic variation that is necessary for a species to adapt to the physical environment and to other organisms arises from new mutations within populations, the recombination of genes during sexual reproduction, and the migration of and interbreeding with individuals from other populations. In very small populations, however, some of that variation is lost by chance alone through random genetic drift. The combined result of these evolutionary processes is that after many generations populations of the same species have widely divergent characteristics. Some of these populations eventually become so genetically different that their members cannot successfully interbreed, resulting in the evolution of a separate species (speciation).

The diversification of life through local adaptation of populations and speciation has created the tremendous biodiversity found on the Earth. In most regions a square kilometre (0.4 square mile) will harbour hundreds—in some places even thousands—of species. The interactions between these species create intricate webs of relationships as the organisms reciprocally evolve, adapting to one another and becoming specialized for their interactions. Natural communities of species reflect the sum of these species' interactions and the ongoing complex selection pressures they constantly endure that drive their evolution. The many ecological and evolutionary processes that affect the relationships among species and their environments render ecology one of the most intricate of the sciences. The answers to the major questions in ecology require an understanding of the relative effects of many variables acting simultaneously.

The Importance of the Biosphere

The continued functioning of the biosphere is dependent not only on the maintenance of the intimate interactions among the myriad species within local communities but also on the looser yet crucial interactions of all species and communities around the globe. The Earth is blanketed with so many species and so many different kinds of biological communities because populations have been able to adapt to almost any kind of environment on Earth through natural selection. Life-forms have evolved that are able to survive in the ocean depths, the frigid conditions of Antarctica, and the near-boiling temperatures of geysers. The great richness of adaptations found among different populations and species of living organisms is the Earth's greatest resource. It is a richness that has evolved over millions of years and is irreplaceable.

It is therefore startling to realize that our inventory of the Earth's diversity is still so incomplete that the total number of living species cannot be estimated more closely than between 3 and 30 million species. Decades of continuous research must be carried out by systematists, ecologists, and geneticists before the inventory of biodiversity provides a more accurate count. The research has been slow. Only recently, as the extinction rate of species has been increasing rapidly, have societies begun to realize the interdependence of species. To sustain life on Earth, more than the few animal and plant species used by humans must be preserved. The flow of energy and the cycling of nutrients through ecosystems, the regulation of populations, and the stability of biological communities, all of which support the continued maintenance of life, rely on the diversity of species, their adaptations to local physical conditions, and their coevolved relationships.

Despite the limited scientific knowledge of most species, ecological studies during the 20th century made great headway in unraveling the mechanisms by which organisms coevolve with one another and adapt to their physical environment, thereby shaping the biosphere. Each new decade has produced a steady stream of studies showing that the biological and physical elements of the Earth are more interconnected than had been previously thought. Those studies also have shown that often the most seemingly insignificant species are crucial to the stability of communities and ecosystems. Many seemingly obscure species are at risk worldwide of being dismissed as unimportant. The effect that the loss of species will have on ecosystems is appreciated only by understanding the relationships between organisms and their environments and by studying the ecological and evolutionary processes operating within ecosystems.

The need to understand how the biosphere functions has never been greater. When human population levels were low and technological abilities crude, societies' impact on the biosphere was relatively small. The increase in human population levels and the harvesting of more of the Earth's natural resources has altered this situation, especially in recent decades. Human activities are causing major alterations to the patterns of energy flow and nutrient cycling through ecosystems, and these activities are eliminating populations and species that have not even been described but which might have been of central importance to the maintenance of ecosystems.

The biologist Edward O. Wilson, who coined the term biodiversity, estimated conservatively that in the late 20th century at least 27,000 species were becoming extinct each year. The majority of these were small tropical organisms. The impact that this freshet of extinctions would have on the biosphere is akin to receiving a box of engine parts and discarding a portion of them before reading the directions, assuming that their absence will have no negative repercussions on the running of the engine.

Resources of the Biosphere

The Photosynthetic Process

Life on Earth depends on the harnessing of solar energy by the process of photosynthesis. Photosynthetic plants convert solar energy into the chemical energy of living tissue, and that stored chemical energy flows into herbivores, predators, parasites, decomposers, and all other forms of life. In the photosynthetic process, light energy is absorbed by the chlorophyll molecules of plants to convert carbon dioxide and water into carbohydrates and oxygen gas. Proteins, fats, nucleic acids, and other compounds also are synthesized during the process, as long as elements such as nitrogen, sulfur, and phosphorus are available.

Efficiency of Solar Energy Utilization

Most solar energy occurs at wavelengths unsuitable for photosynthesis. Between 98 and 99 percent of solar energy reaching the Earth is reflected from leaves and other surfaces and absorbed by other molecules, which convert it to heat. Thus, only 1 to 2 percent is available to be captured by plants. The rate at which plants photosynthesize depends on the amount of light reaching the leaves, the temperature of the environment, and the availability of water and other nutrients such as nitrogen and phosphorus. The measurement of the rate at which organisms convert light energy (or inorganic chemical energy) to the chemical energy of organic compounds is called primary productivity. Hence, the total amount of energy assimilated by plants in an ecosystem during photosynthesis (gross primary productivity) varies among environments. (Productivity is often measured by an increase in biomass, a term used to refer to the weight of all living organisms in an area. Biomass is reported in grams or metric tons.)

Much of the energy assimilated by plants through photosynthesis is not stored as organic material but instead is used during cellular respiration. In this process organic compounds such as carbohydrates, proteins, and fats are broken down, or oxidized, to provide energy (in the form of adenosine triphosphate [ATP]) for the cell's metabolic needs. The energy not used in this process is stored in plant tissues for further use and is called net primary productivity. About 40 to 85 percent of gross primary productivity is not used during respiration and becomes net primary productivity. The highest net primary productivity in terrestrial environments occurs in swamps and marshes and tropical rainforests; the lowest occurs in deserts. In aquatic environments, the highest net productivity occurs in estuaries, algal beds, and reefs. Consequently, these environments are especially critical for the maintenance of worldwide biological productivity.

Energy Transfers and Pyramids

A small amount of the energy stored in plants, between 5 and 25 percent, passes into herbivores (plant eaters) as they feed, and a similarly small percentage of the energy in herbivores then passes into carnivores (animal eaters). The result is a pyramid of energy, with most energy concentrated in the photosynthetic organisms at the bottom of food chains and less energy at each higher trophic level. Some of the remaining energy does not pass directly into the plant-herbivore-carnivore food chain but instead is diverted into the detritus food chain. Bacteria, fungi, scavengers, and carrion eaters that consume detritus (detritivores) are all eventually consumed by other organisms.

The rate at which these consumers convert the chemical energy of their food into their own biomass is called secondary productivity. The efficiency at which energy is transferred from one trophic level to another is called ecological efficiency. On average it is estimated that there is only a 10 percent transfer of energy.

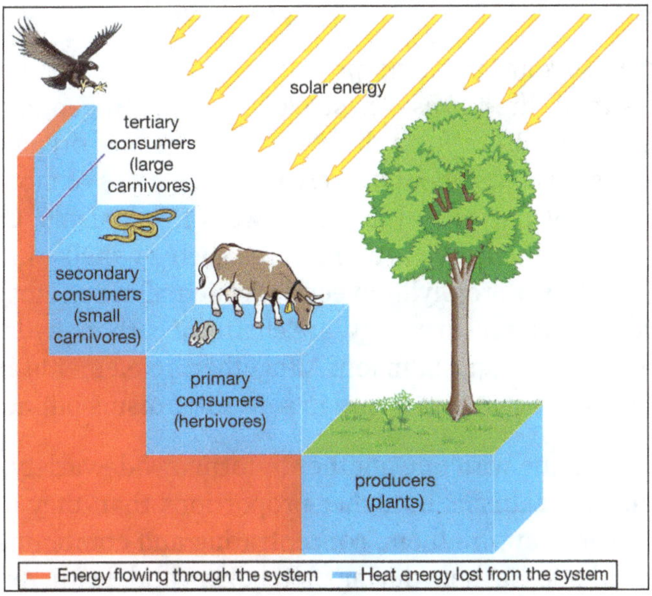

Transfer of energy through an ecosystem. At each trophic level only a small proportion of energy (approximately 10 percent) is transferred to the next level.

Energy is lost in several ways as it flows along these pathways of consumption. Most plant tissue is uneaten by herbivores, and this stored energy is therefore lost to the plant-herbivore-carnivore food chain. In terrestrial communities less than 10 percent of plant tissue is actually consumed by herbivores. The rest falls into the detritus pathway, although the detritivores consume only some of this decaying tissue. Oil and coal deposits are major repositories of this unused plant energy and have accumulated over long periods of geologic time.

The efficiency by which animals convert the food they ingest into energy for growth and reproduction is called assimilation efficiency. Herbivores assimilate between 15 and 80 percent of the plant material they ingest, depending on their physiology and the part of the plant that they eat. For example, herbivores that eat seeds and young vegetation high in energy have the highest assimilation efficiencies, those that eat older leaves have intermediate efficiencies, and those that feed on decaying wood have very low efficiencies. Carnivores generally have higher assimilation efficiencies than herbivores, often between 60 and 90 percent, because their food is more easily digested.

The overall productivity of the biosphere is therefore limited by the rate at which plants convert solar energy (about 1 percent) into chemical energy and the subsequent efficiencies at which other organisms at higher trophic levels convert that stored energy into their own biomass (approximately 10 percent). Human-induced changes in net primary productivity in the parts of the biosphere that have the highest productivity, such as estuaries and tropical moist forests, are likely to have large effects on the overall biological productivity of the Earth.

Nutrient Cycling

The cells of all organisms are made up primarily of six major elements that occur in similar proportions in all life-forms. These elements—hydrogen, oxygen, carbon, nitrogen, phosphorus, and sulfur—form the core protoplasm of organisms, and the first four of these elements make up about 99 percent of the mass of most cells. Additional elements, however, are also essential to the growth of organisms. Calcium and other elements help to form cellular support structures such as shells, internal or external skeletons, and cell walls. Chlorophyll molecules, which allow photosynthetic plants to convert solar energy into chemical energy, are chains of carbon, hydrogen, and oxygen compounds built around a magnesium ion. Altogether, 16 elements are found in all organisms; another eight elements are found in some organisms but not in others.

These bioelements combine with one another to form a wide variety of chemical compounds. They occur in organisms in higher proportions than they do in the environment because organisms capture them, concentrating and combining them in various ways in their cells, and release them during metabolism and death. As a result, these essential nutrients alternate between inorganic and organic states as they rotate through their respective biogeochemical cycles. These cycles can include all or part of the following: the atmosphere, which is made up largely of gases including water vapour; the lithosphere, which encompasses the soil and the entire solid crust of the Earth; and the hydrosphere, which includes lakes, rivers, and oceans.

A portion of the elements are bound up in limestone and in the minerals of other rocks and are unavailable to organisms. The slow processes of weathering and erosion eventually release these elements to enter the cycle. For most of the major nutrients, however, organisms not only intercept the elements moving through the biosphere, but they actually drive the biogeochemical cycles.

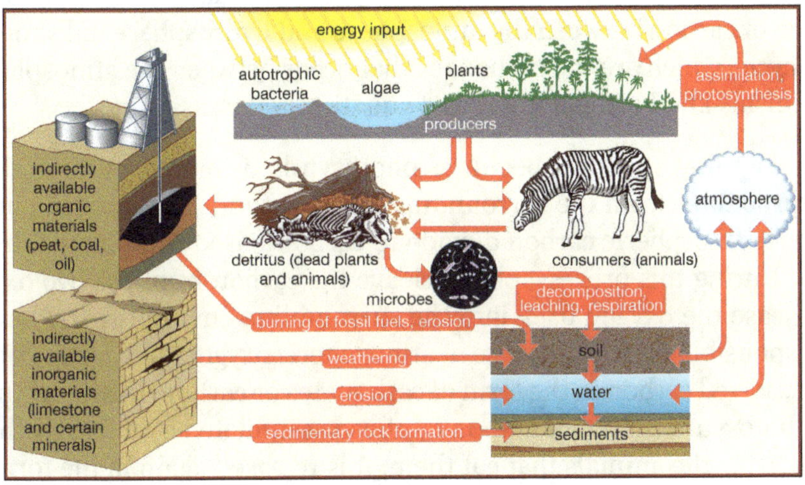

General paths of energy flow in the biosphere.

The movement of nutrients through the biosphere is different from the transfer of energy because, whereas energy flows through the biosphere and cannot be reused, elements are recycled. The same atoms of carbon or nitrogen may, over the course of eons, move repeatedly between organisms, the atmosphere, the soil, and the oceans. Carbon released as carbon dioxide by an animal may remain in the atmosphere for 5 or 10 years before being taken up by another organism, or it may cycle almost immediately back into a neighbouring plant and be used during photosynthesis.

The Carbon Cycle

Life is built on the conversion of carbon dioxide into the carbon-based organic compounds of living organisms. The carbon cycle illustrates the central importance of carbon in the biosphere. Different paths of the carbon cycle recycle the element at varying rates. The slowest part of the cycle involves carbon that resides in sedimentary rocks, where most of the Earth's carbon is stored. When in contact with water that is acidic (pH is low), carbon will dissolve from bedrock; under neutral conditions, carbon will precipitate out as sediment such as calcium carbonate (limestone). This cycling between solution and precipitation is the background against which more rapid parts of the cycle occur.

Short-term cycling of carbon occurs in the continual physical exchange of carbon dioxide (CO_2) between the atmosphere and hydrosphere. Carbon dioxide from the atmosphere becomes dissolved in water (H_2O), with which it reacts to form carbonic acid (H_2CO_3), which dissociates into hydrogen ions (H^+) and bicarbonate ions (HCO_3^-), which further dissociate into hydrogen and carbonate ions (CO_3^{2-}). The more alkaline the water (pH above 7.0 is alkaline), the more carbon is present in the form of carbonate, as is shown in the following reversible reactions:

$$CO_2 + H_2O \rightleftharpoons H_2CO_3 \rightleftharpoons H + HCO_3^- \rightleftharpoons 2H^+ + CO_3^{2-}.$$

At the same time, carbon dioxide in the water is continually lost to the atmosphere. The exchange of carbon between the atmosphere and hydrosphere links the remaining parts of the cycle, which are the exchanges that occur between the atmosphere and terrestrial organisms and between water and aquatic organisms.

The biological cycling of carbon begins as photosynthetic organisms assimilate carbon dioxide or carbonates from the surrounding environment. In terrestrial communities, plants convert atmospheric carbon dioxide to carbon-based compounds through photosynthesis . During this process, plants cleave the carbon from the two oxygen molecules and release the oxygen back into the surrounding environment. Plants are thus primarily responsible for the presence of atmospheric oxygen. In aquatic communities, plants use dissolved carbon in the form of carbonates or carbon dioxide as the source of carbon for photosynthesis. Once carbon has been assimilated by photosynthetic organisms, as well as by the animals that eat them, it is released again in the form of carbon dioxide as these organisms respire. The release of carbon dioxide into the atmosphere or hydrosphere completes the biological part of the carbon cycle.

The pathways of the global carbon cycle, however, are never completely balanced. That is to say, carbon does not move in and out of all parts of the biosphere at equal rates. Consequently, over time some parts of the biosphere accumulate more carbon than others, thereby serving as major accessible carbon reservoirs. In preindustrial times the major reservoirs of carbon were the deep and shallow portions of the ocean; the soil, detritus, and biota of the land; and the atmosphere. The oceans were, and still are, the greatest reservoirs of carbon. Because marine phytoplankton have such short life cycles, the carbon in the ocean cycles rapidly between inorganic and organic states.

In terrestrial environments, forests are the largest carbon reservoirs. Up to 80 percent of the aboveground carbon in terrestrial communities and about a third of belowground carbon are contained within forests. Unlike the oceans, much of this carbon is stored directly in the tissues of plants. High-latitude forests include large amounts of carbon not only in aboveground vegetation but also in peat deposits. Forests at high and low latitudes particularly are important reservoirs of carbon. An estimated one-half of the carbon in forests occurs in high-latitude forests, and a little more than one-third occurs in low-latitude forests. The two largest forest reservoirs of carbon are the vast expanses in Russia, which hold roughly 25 percent of the world's forest carbon, and the Amazon basin, which contains about 20 percent.

Until recent centuries, the equilibrium between the carbon in the world's forests and in the atmosphere remained constant. Samples of carbon dioxide trapped in ice during the past 1,000 years and direct measurements of carbon dioxide in the atmosphere had remained fairly constant until the 18th century. However, the cutting of much of the world's forest is, along with the increase in consumption of fossil fuels attendant on the Industrial Revolution, has resulted in a disruption of the balance between the amount of carbon dioxide in the forests and in the atmosphere. The concentration of atmospheric carbon dioxide has been increasing steadily; currently the rate of increase is about 4 percent per decade. If human activities continue to alter the relative sizes of the carbon reservoirs worldwide, they are likely to have large effects on the carbon cycle and other biogeochemical cycles. Large-scale deforestation in Russia and the Amazon basin are likely to have particularly significant effects on global carbon storage and cycling.

The warming of global temperatures also is changing which ecosystems act as long-term sinks for carbon and which act as sources for carbon dioxide in the atmosphere. For example, the Arctic tundra, with large amounts of carbon stored in its soils, has been a net sink for carbon dioxide during long periods of geologic time. The recent warming of many Arctic regions, however, has accelerated the rate of soil decomposition, transforming these Arctic areas into potential sources of atmospheric carbon dioxide.

The complex cycle of carbon and the varying sizes of carbon reservoirs illustrate some of the reasons it has been so difficult to predict the effects that increased atmospheric carbon will have on global change.

The Nitrogen Cycle

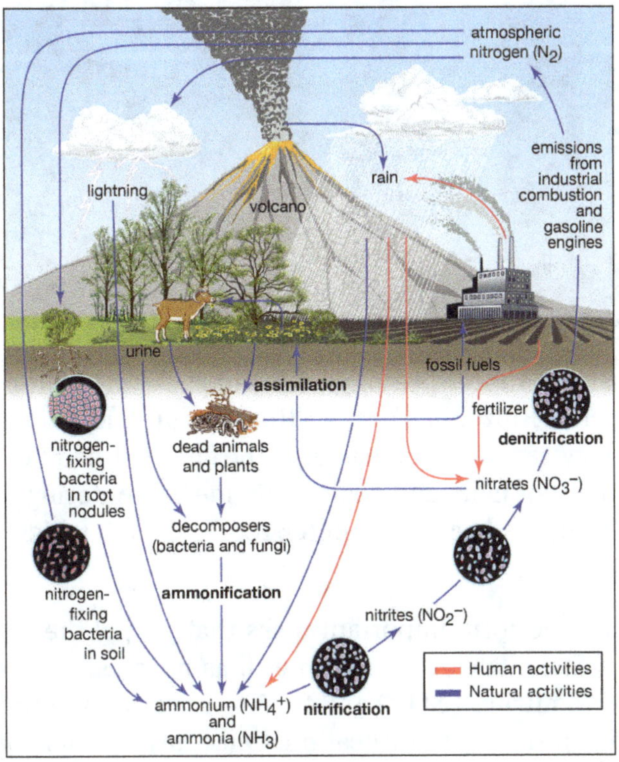

The nitrogen cycle.

Nitrogen is one of the elements most likely to be limiting to plant growth. Like carbon, nitrogen has its own biogeochemical cycle, circulating through the atmosphere, lithosphere, and hydrosphere. Unlike carbon, which is stored primarily in sedimentary rock, most nitrogen occurs in the atmosphere as an inorganic compound (N_2). It is the predominant atmospheric gas, making up about 79 percent of the volume of the atmosphere. Plants, however, cannot use nitrogen in its gaseous form and are able to assimilate it only after it has been converted to ammonia (NH_3) and nitrates (NO_3^-). This reductive process, called nitrogen fixation, is a chemical reaction in which electrons are picked up from another molecule. A small amount of nitrogen is fixed by lightning, but most of the nitrogen harvested from the atmosphere is removed by nitrogen-fixing bacteria and cyanobacteria (formerly called blue-green algae).

Certain species of nitrogen-fixing bacteria can coexist intimately (symbiotically) with legumes and other plants, providing the plants with necessary nitrogen. In this symbiotic association, the bacteria become encased in nodules that grow on the roots of plants, through which nitrogen that has been fixed by the resident bacteria is obtained. Cyanobacteria have developed similar relationships with various life-forms, such as liverworts, hornworts, cycads, and at least one genus of flowering plants (Gunnera). Their symbiotic relationship with fungi has earned its own designation—the coexistent species are called lichen.

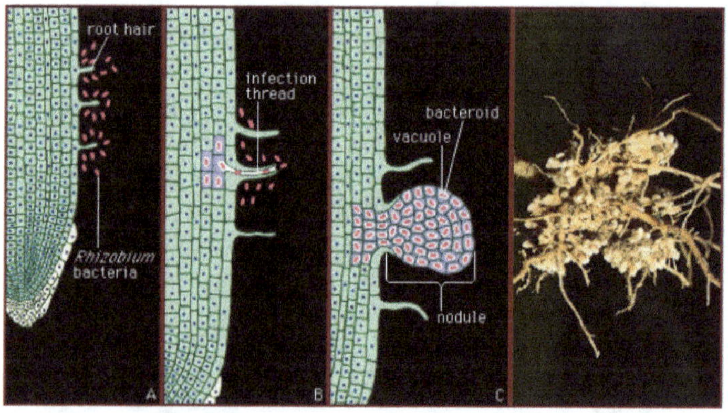

The roots of an Austrian winter pea plant (Pisum sativum) with nodules harbouring nitrogen-fixing bacteria (Rhizobium). (Left) Root nodules develop as a result of a symbiotic relationship between rhizobial bacteria and the root hairs of the plant. (A) The bacteria recognize the root hairs and begin to divide, (B) entering the root through an infection thread that allows bacteria to enter root cells, (C) which divide to form the nodule.

Other microorganisms perform important tasks that propel the nitrogen cycle along. Although plants can assimilate ammonia as well as nitrates, most of the ammonia in the soil is converted to nitrites (NO_2^-) and then to nitrates by certain aerobic bacteria through the oxidative process of nitrification. Once nitrogen has been assimilated by plants, it can be converted to organic forms, such as amino acids and proteins. Animals can use only organic nitrogen, which they obtain by consuming plants or other animals. As these organisms die, certain microbes such as detritivores are able to participate in the decomposition of organic nitrogen into ammonia (ammonification), providing a constant supply of ammonia to be used in the process of nitrification. Although the fixation of atmospheric nitrogen is an essential part of the nitrogen cycle, ammonification and nitrification are the predominant methods by which organic nitrogen is prevented from returning to the atmosphere and is kept cycling through the biosphere.

Some nitrogen does return to the atmosphere, however, as denitrifying bacteria break down nitrates to obtain oxygen, thereby releasing gaseous N_2. Nitrogen is also lost from plants and soil in terrestrial environments via other routes, including erosion, runoff, volatilization of ammonia into the atmosphere, and leaching from soils into lakes and streams. Eventually some of these nutrients reach the oceans as rivers flush them onto the ocean surface.

The Sulfur Cycle

Sulfur is found in all living organisms as a constituent of some proteins, vitamins, and hormones. Like carbon and nitrogen, sulfur cycles between the atmosphere, lithosphere, and hydrosphere; but, unlike these two other elements, it has major reservoirs in both the atmosphere and the lithosphere. As is true in the nitrogen cycle, the activities of microorganisms are crucial in the global cycling of this nutrient.

The process begins with geochemical and meteorologic processes such as the weathering of rock. When sulfur is released from the rock and comes in contact with air, it is converted into sulfate (SO_4), which is taken up by plants and microorganisms and converted into organic forms. Animals acquire these organic forms of sulfur from their foods. When organisms die and decompose, some of the sulfur enters the tissues of microorganisms and some is released again as sulfate. There is, however, a continual loss of sulfur from terrestrial ecosystems as some of it drains into lakes and streams and eventually into the ocean as runoff. Additional sulfur enters the ocean through fallout from the atmosphere.

Once in the ocean, some of the sulfur cycles through marine communities as it moves through food chains, some reenters the atmosphere, and some is lost to the ocean depths as it combines with iron to form ferrous sulfide (FeS), which is responsible for the black colour of marine sediments. Sulfur reenters the atmosphere naturally in three major ways: sea spray releases large amounts of the element from the ocean into the atmosphere; anaerobic respiration by sulfate-reducing bacteria causes the release of hydrogen sulfide (H_2S) gas especially from marshes, tidal flats, and similar environments in which anaerobic microorganisms thrive; and volcanic activity releases additional but much smaller amounts of sulfur gas into the atmosphere.

Since the Industrial Revolution, human activities have contributed significantly to the movement of sulfur from the lithosphere to the atmosphere as the burning of fossil fuels and the processing of metals have occasioned large emissions of sulfur dioxide.

Oxides of sulfur and nitrogen contribute to the acid rain that is common downwind from these industrial activities.

The Cycling of Phosphorus and other Essential Nutrients

Most other major nutrients such as phosphorus, potassium, magnesium, iron, and calcium enter terrestrial communities through the weathering of bedrock. These nutrients lack a volatile gaseous state. Consequently, they cycle through the biosphere differently from carbon, nitrogen, and sulfur, all of which sometimes occur as volatile gases. Of the nonvolatile nutrients, phosphorus is the one that most often limits plant growth, especially in aquatic environments.

Phosphorus and the other non-volatile elements move unidirectionally from land, through aquatic environments, into ocean sediments. Most phosphorus cycling occurs between the surface and depths of the ocean. When near the surface, phosphorus is taken up by the plankton and passed through the food chain. It cycles back to the ocean bottom as individuals die and fall to the ocean floor, releasing assimilated phosphorus. Much of this element gradually accumulates in the ocean sediment as particulate phosphorus and is eventually brought back to the surface only through ocean upwelling and tectonic activity. The ocean sediments are therefore by far the greatest reservoirs of phosphorus.

In terrestrial ecosystems, much of the available phosphorus moves in a closed cycle between living organisms and the organic debris in the soil. Phosphate (PO_4^{3-}) is the only important inorganic form involved in this cycle. Microorganisms in the soil break down litter and other organic matter, thereby releasing phosphate, which is then taken up by plants and released again when they die and decompose. Soils differ in the amount of phosphorus they contain, and in some phosphorus-poor soils almost all the available phosphorus resides in living organisms and organic debris. In some tropical forests, such as those in parts of Brazil, so much of the phosphorus is contained in living organisms that the clearing of forests eliminates most of this element. As a result, the plant communities cannot recover, and crops cannot be grown.

The addition of phosphorus to soils poor in this nutrient and the use of phosphorus-rich detergents have had a great impact on the phosphorus cycle in many ecosystems. Runoff from agricultural fields and the emptying of sewage has introduced so much extra phosphorus to rivers and lakes that populations of plants and microorganisms have grown explosively, sometimes creating a solid mat that extends over the surface of the water. This growth increases the amount of organic debris in the water, which can lead to a decrease in the available oxygen, resulting in suffocation of fish and other animals.

The Hydrologic Cycle

A portion of the biogeochemical cycle of all elements involves time spent cycling through the hydrosphere. Water itself cycles within the biosphere. Unlike the cycles of the other major nutrients, however, the hydrologic, or water, cycle would continue in some form even in the absence of living organisms. Most of the Earth's water is in its core, in the sedimentary rocks near its surface, and in the ocean. A minute percentage of the water, however, continually cycles through the atmosphere, oceans, and terrestrial environments mainly by the processes of evaporation and precipitation.

This part of the hydrologic cycle is driven by solar energy. Water evaporates from both the aquatic and terrestrial environments as it is heated by the Sun's energy. The rates of evaporation and precipitation depend on solar energy, as do the patterns of circulation of moisture in the air and currents in the ocean. Evaporation exceeds precipitation over the oceans, and this water vapour is transported by the wind over land, where it returns to the land through precipitation. The water falling onto terrestrial environments seeps into the ground or runs off into lakes and streams and eventually empties into the oceans, carrying with it many of the other major nutrients. Water also re-enters the atmosphere through the evaporative loss of water by plants (transpiration).

Links among the Cycles

Although the overall pattern of cycling of all the major elements is now known, there is still much to learn about the relative importance of the different stages of each cycle. For example, there is considerable debate concerning which ecosystems act as the major sources of carbon for the atmosphere and which act as sinks by accumulating more carbon than they release. The ways in which the different cycles interact with one another also must be minutely studied. It has been discovered that sulfur availability influences the rate of nitrogen accumulation in plants and nitrogen availability influences phosphorus uptake. All three elements are thought to influence the rate of carbon accumulation by plants. As a result, changes in any one of these nutrient cycles influence the other cycles as well.

The effects that disruptions in these cycles may have within the biosphere are not clearly understood. Natural geologic phenomena, such as ice ages and major periods of volcanic activity, have repeatedly disturbed these cycles throughout Earth history. Many human activities may have impacts of similar scope. Deforestation, the burning of fossil fuels, and increased fertilization are disturbing these cycles. These anthropogenic disturbances have increased atmospheric levels of carbon dioxide, decreased ozone (O_3) levels, and undermined the natural equilibrium of streams and lakes by excessive nutrient enrichment from sewage, fertilizers, and factory waste (cultural eutrophication). Gleaning more information about the biogeochemical cycles and their interactions has become increasingly important now that the effects of human activities are becoming more apparent.

Another potential effect that may result from human intrusions in the environment is global warming. Carbon dioxide in the atmosphere has the ability to act as an insulator, preventing some of the Earth's heat, absorbed from solar radiation, from escaping back into space. This process, known as the greenhouse effect, is suspected of being enhanced by rising levels of atmospheric carbon dioxide, which have resulted in part from the combustion of fossil fuels and the clearing and burning of tropical forests. This increase in atmospheric carbon dioxide and other so called greenhouse gases could raise the overall global temperature, causing the polar ice caps to melt, sea levels to rise, and the Earth's precipitation to be redistributed.

The complexity and interconnectedness of each of the biogeochemical cycles make it difficult to pinpoint how any one human activity is altering the cycles; nevertheless, the majority of those who study these fluctuations agree that this is happening. Disagreements generally concern the extent to which various activities affect particular cycles and what the long-term consequences of these disturbances will be.

Environmental Conditions

Most organisms are limited to either a terrestrial or an aquatic environment. An organism's ability to tolerate local conditions within its environment further restricts its distribution. One parameter, such as temperature tolerance, may be important in determining the limits of distribution, but often a combination of variables, such as temperature tolerance and water requirements, is important. Extreme environmental variables can evoke physiological and behavioral responses from organisms. The physiological response helps the organism maintain a constant internal environment (homeostasis), while a behavioral response allows it to avoid the environmental challenge—a fallback strategy if homeostasis cannot be maintained.

The ways in which modern living organisms tolerate environmental conditions reflect the aquatic origins of life. With few exceptions, life cannot exist outside the temperature range at which water is a liquid. Thus, liquid water, and temperatures that maintain water as a liquid, are essential for sustaining life. Within those parameters, the concentrations of dissolved salts and other ions, the abundance of respiratory gases, atmospheric or hydrostatic pressure, and rate of water flow all influence the physiology, behaviour, and distribution of organisms.

Temperature

Temperature has the single most important influence on the distribution of organisms because it determines the physical state of water. Most organisms cannot live in conditions in which the temperature remains below 0 °C or above 45 °C for any length of time. Adaptations have enabled certain species to survive outside this range thermophilic bacteria have been found in hot springs in which the temperatures may approach the boiling point, and certain polar mosses and lichens can tolerate temperatures of −70 °C but these species are the exceptions. Few organisms can remain for long periods at temperatures above 45 °C, because organic molecules such as proteins will begin to denature. Nor are temperatures below freezing conducive to life: cells will rupture if the water they contain freezes.

Most organisms are not able to maintain a body temperature that is significantly different from that of the environment. Sessile organisms, such as plants and fungi, and very small organisms and animals that cannot move great distances, therefore, must be able to withstand the full range of temperatures sustained by their habitat. In contrast, many mobile animals employ behavioral mechanisms to avoid extreme conditions in the short term. Such behaviours vary from simply moving short distances out of the Sun or an icy wind to large-scale migrations.

Some types of animals employ physiological mechanisms to maintain a constant body temperature, and two categories are commonly distinguished: the term cold-blooded is understood to refer to reptiles and invertebrates, and warm-blooded is generally applied to mammals and birds. These terms, however, are imprecise; the more accurate terms, ectotherm for cold-blooded and endotherm for warm-blooded, are more useful in describing the thermal capabilities of these animals. Ectotherms rely on external sources of heat to regulate their body temperatures, and endotherms thermoregulate by generating heat internally.

Terrestrial ectotherms utilize the complex temperature profile of the terrestrial environment to derive warmth. They can absorb solar radiation, thus raising their body temperatures above that of the surrounding air and substrate, unlike the aquatic ectotherm, whose body temperature is usually very close to that of the environment. As this solar radiation is taken up, physiological mechanisms contribute to the regulation of heat—peripheral blood vessels dilate and heart rate increases. The animal also may employ behavioral mechanisms, such as reorienting itself toward the Sun or flattening its body and spreading its legs to maximize its surface area exposure. At night, loss of heat may be reduced by other behavioral and physiological mechanisms—the heart rate may slow, peripheral blood vessels may constrict, surface area may be minimized, and shelter may be sought.

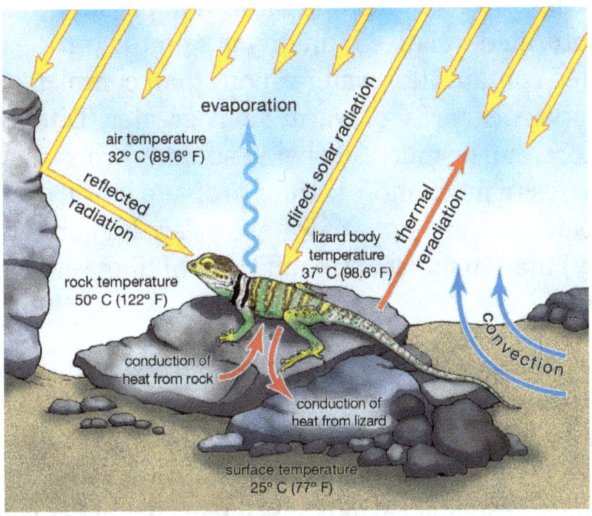

Energy exchange between a terrestrial reptile and the environment.

Endotherms maintain body temperature independently of the environment by the metabolic production of heat. They generate heat internally and control passive heat loss by varying the quality of their insulation or by repositioning themselves to alter their effective surface area (i.e., curling into a tight ball). If heat loss exceeds heat generation, metabolism increases to make up the loss. If heat generation exceeds the rate of loss, mechanisms to increase heat loss by evaporation occur. In either case, behavioral mechanisms can be employed to seek a more suitable thermal environment.

To survive for a limited period in adverse conditions, endotherms may employ a combination of behavioral and physiological mechanisms. In cold weather, which requires an increase in energy consumption, the animal may enter a state of torpor in which its body temperature, metabolism, respiratory rate, and heart rate are depressed. Long-term winter hypothermia, or hibernation, is an extended state of torpor that some animals use as a response to cold conditions. Torpor and hibernation free the animals from energetically expensive maintenance of high body temperatures, saving energy when food is limited.

Another form of torpor, estivation, is experienced by animals in response to heat stress. This state is seen more often in ectothermic animals than in endotherms, but in both the stimulus for estivation is usually a combination of high temperatures and water shortage.

Humidity

Most terrestrial organisms must maintain their water content within fairly narrow limits. Water commonly is lost to the air through evaporation or, in plants, transpiration. Because most water loss occurs by diffusion and the rate of diffusion is determined by the gradient across the diffusion barrier such as the surface of a leaf or skin, the rate of water loss will depend on the relative humidity of the air. Relative humidity is the percent saturation of air relative to its total saturation possible at a given temperature. When air is totally saturated, relative humidity is said to be 100 percent. Cool air that is completely saturated contains less water vapour than completely saturated warm air because the water vapour capacity of warm air is greater. Diffusion gradients across skin or leaves, therefore, can be much steeper in summer when the air is warm, rendering evaporative water loss a much more serious problem in warm environments than in cool environments. Nevertheless, rates of water loss are higher in dry air (conditions of low relative humidity) than in moist air (conditions of high relative humidity), regardless of the temperature.

Water loss from evaporation must be compensated by water uptake from the environment. For most plants, transpirational water loss is countered by the uptake of water from the soil via roots. For animals, water content can be replenished by eating or drinking or by uptake through the integument. For organisms living in dry environments, there are many morphological and physiological mechanisms that reduce water loss. Desert plants, or xerophytes, typically have reduced leaf surface areas because leaves are the major sites of transpiration. Some xerophytes shed their leaves altogether in summer, and some are dormant during the dry season.

Desert animals typically have skin that is relatively impervious to water. The major site of evaporation is the respiratory exchange surface, which must be moist to allow the gaseous exchange of oxygen and carbon dioxide. A reduction in amount of water lost through respiration can occur if the temperature of the exhaled air is lower than the temperature of the body. As many animals, such as gazelles, inhale warm air, heat and water vapour from the nasal passages evaporate, cooling the nose and the blood within it. The cool venous blood passes close to and cools the warm arterial blood traveling to the brain. If the brain does not require cooling, the venous blood returns to the heart by another route. The nasal passages also cool the warm, saturated air from the lungs so that water condenses in the nose and is reabsorbed rather than lost to the environment.

pH

The relative acidity or alkalinity of a solution is reported by the pH scale, which is a measure of the concentration of hydrogen ions in solution. Neutral solutions have a pH of 7. A pH of less than 7 denotes acidity (an increased hydrogen ion concentration), and above 7 alkalinity (a decreased hydrogen ion concentration). Many important molecular processes within the cells of organisms occur within a very narrow range of pH. Thus, maintenance of internal pH by homeostatic mechanisms is vital for cells to function properly. Although pH may differ locally within an organism, most tissues are within one pH unit of neutral. Because aquatic organisms generally have somewhat permeable skins or respiratory exchange surfaces, external conditions can influence internal pH. These organisms may accomplish the extremely important task of regulating internal pH by exchanging hydrogen ions for other ions, such as sodium or bicarbonate, with the environment.

The pH of naturally occurring waters can range from very acidic conditions of about 3 in peat swamps to very alkaline conditions of about 9 in alkaline lakes. Naturally acidic water may result from the presence of organic acids, as is the case in a peat swamp, or from geologic conditions such as sulfur deposits associated with volcanic activity. Naturally occurring alkaline waters usually result from inorganic sources. Most organisms are unable to live in conditions of extreme alkalinity or acidity.

Salinity

The term salinity refers to the amount of dissolved salts that are present in water. Sodium and chloride are the predominant ions in seawater, and the concentrations of magnesium, calcium, and sulfate ions are also substantial. Naturally occurring waters vary in salinity from the almost pure water, devoid of salts, in snowmelt to the saturated solutions in salt lakes such as the Dead Sea. Salinity in the oceans is constant but is more variable along the coast where seawater is diluted with freshwater from runoff or from the emptying of rivers. This brackish water forms a barrier separating marine and freshwater organisms.

The cells of organisms also contain solutions of dissolved ions, but the range of salinity that occurs in tissues is more narrow than the range that occurs in nature. Although a minimum number of ions must be present in the cytoplasm for the cell to function properly, excessive concentrations of ions will impair cellular functioning. Organisms that live in aquatic environments and whose integument is permeable to water, therefore, must be able to contend with osmotic pressure. This pressure arises if two solutions of unequal solute concentration exist on either side of a semipermeable membrane such as the skin. Water from the solution with a lower solute concentration will cross the membrane diluting the more highly concentrated solution until both concentrations are equalized. If the salt concentration of an animal's body fluids is higher than that of the surrounding environment, the osmotic pressure will cause water to diffuse through the skin until the concentrations are equal unless some mechanism prevents this from happening.

Many marine invertebrates have the same osmotic pressure as seawater. When the salt concentration of their surroundings changes, however, they must be able to adjust. Two means of contending with this situation are employed, and, depending on how they regulate the salt concentrations of their tissues, organisms are classified as osmoregulators or osmoconformers. The osmotic concentration of the body fluids of an osmoconformer changes to match that of its external environment, whereas an osmoregulator controls the osmotic concentration of its body fluids, keeping them constant in spite of external alterations. Aquatic organisms that can tolerate a wide range of external ion concentrations are called euryhaline; those that have a limited tolerance are called stenohaline.

Even if aquatic organisms have an integument that is relatively impermeable to water, as well as to small inorganic ions, their respiratory exchange surfaces are permeable. Hence, organisms occurring in water that has a lower solute concentration than their tissues (e.g., trout in mountain streams) will constantly lose ions to the environment as water flows into their tissues. In contrast, organisms in salty environments face a constant loss of water and an influx of ions.

Many mechanisms have evolved that deal with these problems. Because water cannot be readily pumped across cell membranes, salinity balance is usually maintained by actively transporting inorganic ions, usually sodium and chloride. This process consumes energy and can usurp a large portion of the energy budget of animals in very saline environments. In marine fish, gill cells pump ions out of the body into the sea, while in freshwater fish gill cells pump ions in the opposite direction. Passive water loss in marine fish is compensated primarily in one of two ways. Most bony fish drink copiously and excrete salt across the gills, while the majority of sharks artificially elevate the salt concentration of their tissues above that of seawater with urea and other organic molecules, allowing water to slowly and passively enter the body. Through their food and across their gills, freshwater fish replenish most of the ions they lose. They also produce large quantities of very dilute urine to excrete excess water that diffuses into their bodies.

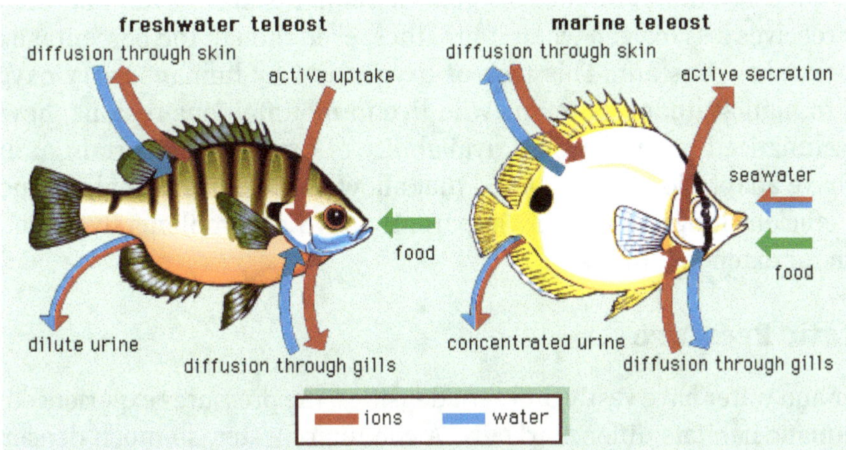

Osmotic regulation in freshwater and marine teleost fish.

Water Currents

The flow of water presents special problems for aquatic organisms. Flow is associated with rivers, oceanic currents, and waves and can be laminar (streamlined) or turbulent. Many organisms are specialized to live in flowing environments; the main obstacle to this lifestyle is the constant threat of being washed away. Both plants and animals have evolved mechanisms that help to anchor them to the substratum in flowing water (e.g., the holdfast of kelp or the byssus threads of mussels). If anchorage can be assured, there are many advantages to living in this environment. Flowing water generally is well oxygenated, and the supply is continuous; nutrients and food are constantly replenished as well. The very precariousness of the environment also affords some protection from predation because the number of predators that make this type of habitat home is limited.

Pressure

Atmospheric Pressure

Variations in atmospheric pressure can present special problems for the respiratory systems of animals because atmospheric pressure affects the exchange of oxygen and carbon dioxide that occurs during animal respiration. Normal atmospheric pressure at sea level is the total pressure that a column of air above the surface of the Earth exerts (760 millimetres of mercury, or 1 atmosphere). The total pressure is the sum of the pressures that each gas—mainly nitrogen, oxygen, and carbon dioxide—would exert alone (the partial pressure of that gas: The gases in the environment). As an animal breathes, oxygen moves from the environment across the respiratory surfaces into the blood; carbon dioxide moves in the reverse direction. This process occurs primarily by passive diffusion; each gas moves from an area of greater to lesser partial pressure, driven by the differential that exists across the respiratory surface. At higher altitudes, where the atmospheric pressure is lower, the partial pressure of oxygen is also lower. The partial pressure differential of oxygen, therefore, is also lower, and the organism effectively receives less oxygen when it breathes, even though the percentage of oxygen in the air remains constant. This lack of oxygen is why humans carry oxygen when ascending to high altitudes. Humans who live in mountainous regions, however, can become acclimatized to the lowered availability of oxygen, and certain animals such as llamas have adaptations of the blood that allow them to live at high altitudes. Birds have very efficient lungs, and many apparently have no problems flying to high altitudes, even for extended flights.

Hydrostatic Pressure

Because air and water have vastly different densities, the pressures experienced in terrestrial and aquatic habitats differ markedly. A column of water, so much denser than air, exerts a greater amount of pressure than a column of air. With each 10-metre (32.8-foot) increase in depth, there is an increase in hydrostatic pressure equivalent to one atmosphere. Mean ocean depth is about 3,800 metres and has a pressure of about 380 atmospheres. To surmount this environmental challenge, animals that live at great depths lack air compartments such as lungs or swim bladders. Surface-dwelling animals that dive to great depths meet this challenge differently. As pressure increases during a dive, air compartments compress, returning to their former volume when the animal surfaces. Air is forced into the trachea, bronchi, and bronchioles, where no gas uptake occurs. Thus, the increased pressure cannot drive more gases into the bloodstream, and, as the animal rises, it does not experience the "bends" (decompression sickness resulting from a rapid reduction of air pressure). In contrast, sea snakes avoid the bends by excreting nitrogen across the skin to offset the uptake of this gas from the lungs.

LITHOSPHERE

The lithosphere is the solid, outer part of the Earth. The lithosphere includes the brittle upper portion of the mantle and the crust, the outermost layers of Earth's structure. It is bounded by the atmosphere above and the asthenosphere (another part of the upper mantle).

Although the rocks of the lithosphere are still considered elastic, they are not viscous. The asthenosphere is viscous, and the Lithosphere-Asthenosphere Boundary (LAB) is the point where geologists and rheologists—scientists who study the flow of matter—mark the difference in ductility between the two layers of the upper mantle. Ductility measures a solid material's ability to deform or stretch under stress. The lithosphere is far less ductile than the asthenosphere.

There are two types of lithosphere: oceanic lithosphere and continental lithosphere. Oceanic lithosphere is associated with oceanic crust, and is slightly denser than continental lithosphere.

Plate Tectonics

The most well-known feature associated with Earth's lithosphere is tectonic activity. Tectonic activity describes the interaction of the huge slabs of lithosphere called tectonic plates.

The lithosphere is divided into tectonic plates including the North American, Caribbean, South American, Scotia, Antarctic, Eurasian, Arabian, African, Indian, Philippine, Australian, Pacific, Juan de Fuca, Cocos, and Nazca.

Most tectonic activity takes place at the boundaries of these plates, where they may collide, tear apart, or slide against each other. The movement of tectonic plates is made possible by thermal energy (heat) from the mantle part of the lithosphere. Thermal energy makes the rocks of the lithosphere more elastic.

Tectonic activity is responsible for some of Earth's most dramatic geologic events: earthquakes, volcanoes, orogeny (mountain-building), and deep ocean trenches can all be formed by tectonic activity in the lithosphere.

Tectonic activity can shape the lithosphere itself: Both oceanic and continental lithospheres are thinnest at rift valleys and ocean ridges, where tectonic plates are shifting apart from one another.

Lithosphere Interacting with other Spheres

The cool, brittle lithosphere is just one of five great "spheres" that shape the environment of Earth. The other spheres are the biosphere (Earth's living things); the cryosphere (Earth's frozen regions, including both ice and frozen soil); the hydrosphere (Earth's liquid water); and the atmosphere (the air surrounding our planet). These spheres interact to influence such diverse elements as ocean salinity, biodiversity, and landscape.

For instance, the pedosphere is part of the lithosphere made of soil and dirt. The pedosphere is created by the interaction of the lithosphere, atmosphere, cryosphere, hydrosphere, and biosphere. Enormous, hard rocks of the lithosphere may be ground down to powder by the powerful movement of a glacier (cyrosphere). Weathering and erosion caused by wind (atmosphere) or rain (hydrosphere) may also wear down rocks in the lithosphere. The organic components of the biosphere, including plant and animal remains, mix with these eroded rocks to create fertile soil—the pedosphere.

The lithosphere also interacts with the atmosphere, hydrosphere, and cryosphere to influence temperature differences on Earth. Tall mountains, for example, often have dramatically lower temperatures than valleys or hills. The mountain range of the lithosphere is interacting with the lower air pressure of the atmosphere and the snowy precipitation of the hydrosphere to create a cool or even icy climate zone. A region's climate zone, in turn, influences adaptations necessary for organisms of the region's biosphere.

The rocky lithosphere includes part of the upper mantle and crust.

GEOSPHERE

The geosphere is considered that portion of the Earth system that includes the Earth's interior, rocks and minerals, landforms and the processes that shape the Earth's surface. The Earth itself is not a perfect sphere. It is what is called an oblate spheroid, with a radius of 6,357 kilometers (km) from the Earth's center to the North Pole and 6,378 km from the center to the Equator. Prior to advanced instruments and spacecraft, 17th-century scientist Sir Isaac Newton predicted a similar shape based on the effects of the Earth's daily rotation and his studies of other planets. Geodesy (the study of the

Earth's shape) is a very important science, in that it is critical for helping us understand satellite orbits, create maps and navigate on the planet using devices such as the Global Positioning System (GPS).

The Earth's interior includes a thin, 5 to 70 km thick layer of oceanic and continental crust overlying an additional 6,300 km of rock and metals. The crust varies in thickness and density, with oceanic crust consisting of a thin (around 5 km) layer of dense rock and continental crust consisting of less-dense, lighter-colored rock ranging between 30 and 70 km in thickness. Although the crust is comprised of many types of rocks and hundreds of minerals, these materials are assembled from a very small number of elements. A total of 98.7% of the crust (by weight) consists of just 8 elements, including oxygen (46.6%), silicon (27.72%), aluminum (8.13%), iron (5.00%), calcium (3.63%), sodium (2.83%), potassium (2.70%) and magnesium (2.09%). These elements form the building blocks of most of the inorganic materials we encounter in our daily lives such as glass (SiO_2), concrete ($CaCO_3$), and steel. Oceanic crust is dominated by minerals consisting of silicon, oxygen and magnesium and is thus called SIMA; continental crust is made up of SIAL, in which silicon and aluminum dominate.

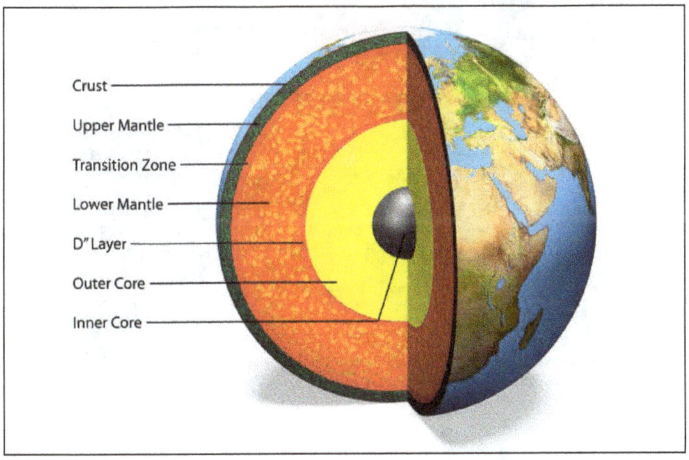

The Earth's interior is arranged somewhat like a layer cake, consisting of a series of layers that change in density, mineral composition and thickness with depth. Directly below the crust is the mantle. It consists of two parts, an upper layer that is less dense and relatively brittle and a lower (much thicker) layer that is more dense and plastic (it deforms without breaking). The crust and upper mantle combined form the brittle upper layers of the Earth's interior called the lithosphere. The upper mantle is also called the asthenosphere.

The mantle makes up the largest volume of the Earth's interior. The region beneath the mantle is called the core, and consists of two parts, a liquid outer core that is around 2250 km thick and a solid inner core 1220 km thick. The core is primarily made up of iron, with a small amount of nickel. The liquid iron in the outer core is particularly important in that it is the primary source of the Earth's magnetic field. Unlike a common magnet, though, the north and south ends of our "global magnet" are not exactly situated at Earth's poles. Instead, the magnetic north pole is actually situated in northern Canada, and the magnetic south pole resides north of Antarctica and south of Australia. Another interesting feature of the magnetic poles is that their precise location moves over time. Every few million years, even the polarity of the Earth's magnetic field reverses (called a geomagnetic reversal, where magnetic north and south "switch"). While scientists still do not fully understand why geomagnetic reversals occur, the presence of changing magnetic orientations preserved in rocks containing iron was a fundamental clue in unravelling the puzzle of Plate Tectonics. Almost all of our direct knowledge of the Earth's interior is from the upper 10 km. Our knowledge of the remaining 6,300 km is based largely on indirect evidence from seismology, laboratory studies of igneous and metamorphic rocks, computer models and meteorites.

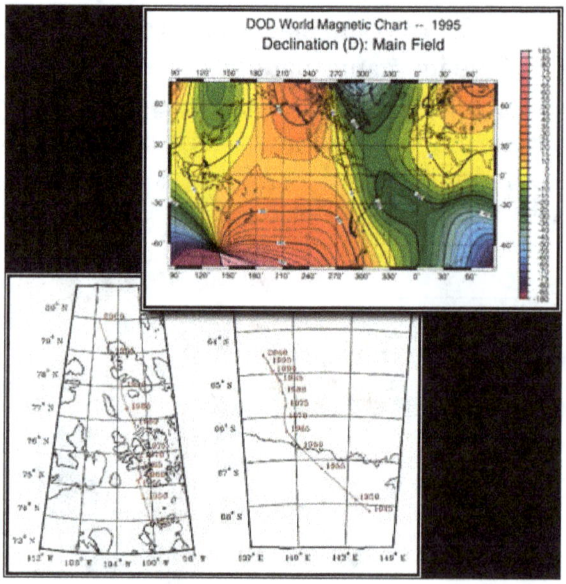

HYDROSPHERE

Hydrosphere is a discontinuous layer of water at or near Earth's surface. It includes all liquid and frozen surface waters, groundwater held in soil and rock, and atmospheric water vapour.

Water is the most abundant substance at the surface of Earth. About 1.4 billion cubic km (326 million cubic miles) of water in liquid and frozen form make up the oceans, lakes, streams, glaciers, and groundwaters found there. It is this enormous volume of water, in its various manifestations, that forms the discontinuous layer, enclosing much of the terrestrial surface, known as the hydrosphere.

Central to any discussion of the hydrosphere is the concept of the water cycle (or hydrologic cycle). This cycle consists of a group of reservoirs containing water, the processes by which water is transferred from one reservoir to another (or transformed from one state to another), and the rates of transfer associated with such processes. These transfer paths penetrate the entire hydrosphere, extending upward to about 15 km (9 miles) in Earth's atmosphere and downward to depths on the order of 5 km (3 miles) in its crust.

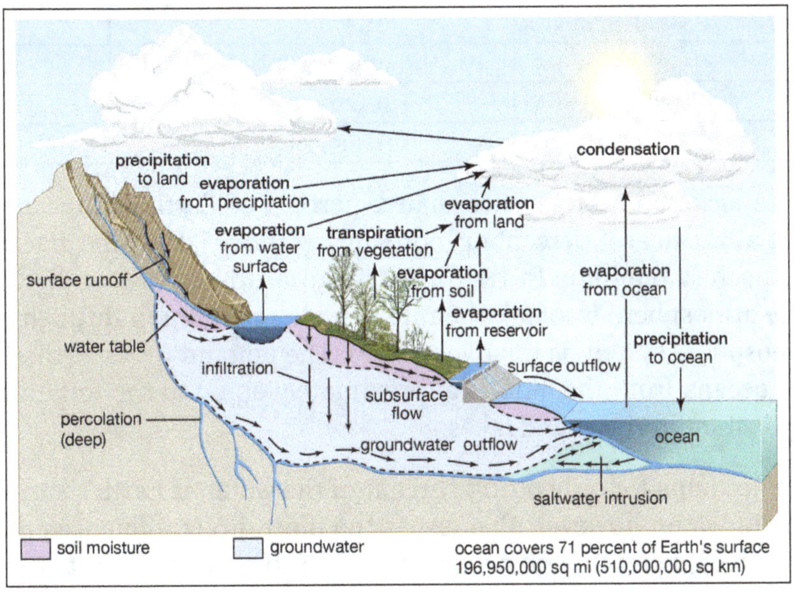

In the hydrologic cycle, water is transferred between the land surface, the ocean, and the atmosphere.

Distribution and Quantity of Earth's Waters

Ocean waters and waters trapped in the pore spaces of sediments make up most of the present-day hydrosphere. The total mass of water in the oceans equals about 50 percent of the mass of sedimentary rocks now in existence and about 5 percent of the mass of Earth's crust as a whole. Deep and shallow groundwaters constitute a small percentage of the total water locked in the pores of sedimentary rocks—on the order of 3 to 15 percent. The amount of water in the atmosphere at any one time is trivial, equivalent to roughly 13,000 cubic km (about 3,100 cubic miles) of liquid water, or about 0.001 percent of the total at Earth's surface. This water, however, plays an important role in the water cycle.

Water masses at earth's surface		
Reservoir	Volume (in cubic kilometres)	Percent of total
Oceans	1,338,000,000	96.5
Ice caps, glaciers, and permanent snow	24,064,000	1.74
Ground ice and permafrost	300,000	0.22
Groundwater (total)	23,400,000	1.69
Groundwater (fresh)	10,530,000	0.76
Groundwater (saline)	12,870,000	0.93
Lakes (total)	176,400	0.013

Lakes (fresh)	91,000	0.007
Lakes (saline)	85,400	0.006
Soil moisture	16,500	0.001
Atmosphere	12,900	0.001
Swamp water	11,470	0.0008
Rivers	2,120	0.0002
Biota	1,120	0.0001
Total	1,409,560,910	101.67

At present, ice locks up a little more than 2 percent of Earth's water and may have accounted for as much as 3 percent or more during the height of the glaciations of the Pleistocene Epoch (2.6 million to 11,700 years ago). Although water storage in rivers, lakes, and the atmosphere is small, the rate of water circulation through the rain-river-ocean-atmosphere system is relatively rapid. The amount of water discharged each year into the oceans from the land is approximately equal to the total mass of water stored at any instant in rivers and lakes.

Soil moisture accounts for only 0.005 percent of the water at Earth's surface. It is this small amount of water, however, that exerts the most direct influence on evaporation from soils. The biosphere, though primarily H_2O in composition, contains very little of the total water at the terrestrial surface, only about 0.00004 percent, yet the biosphere plays a major role in the transport of water vapour back into the atmosphere by the process of transpiration.

Earth's waters are not pure H_2O but contain dissolved and particulate materials. Thus, the masses of water at Earth's surface are major receptacles of inorganic and organic substances, and water movement plays a dominant role in the transportation of these substances about the planet's surface.

Biogeochemical Properties of the Hydrosphere

Rainwater

About 107,000 cubic km (nearly 25,800 cubic miles) of rain fall on land each year. The total water in the atmosphere is 13,000 cubic km, and this water, owing to precipitation and evaporation, turns over every 9.6 days. Rainwater is not pure but rather contains dissolved gases and salts, fine-ground particulate material, organic substances, and even bacteria. The sources of the materials in rainwater are the oceans, soils, fertilizers, air pollution, and fossil fuel combustion.

It has been observed that rains over oceanic islands and near coasts have ratios of major dissolved constituents very close to those found in seawater. The discovery of the high salt content of rain near coastlines was somewhat surprising because sea salts are not volatile, and it might be expected that the process of evaporation of water from the sea surface would "filter" out the salts. It has been demonstrated, however, that a large percentage of the salts in rain is derived from the bursting of small bubbles at the sea surface due to the impact of rain droplets or the breaking of waves, which results in the injection of sea aerosol into the atmosphere. This sea aerosol evaporates, with resultant precipitation of the salts as tiny particles that are subsequently carried high into the atmosphere by turbulent winds. These particles may then be transported over continents to fall in rain or as dry deposition.

Assuming equilibrium with the atmospheric carbon dioxide partial pressure (PCO_2) of 10–3.5 (0.00035) atmosphere, the approximate mean composition of rainwater is in parts per million (ppm): sodium (Na^+), 1.98; potassium (K^+), 0.30; magnesium (Mg_2^+), 0.27; calcium (Ca_2^+), 0.09; chloride (Cl^-), 3.79; sulfate (SO_4^{2-}), 0.58; and bicarbonate (HCO_3^-), 0.12. In addition to these ions, rainwater contains small amounts of dissolved silica—about 0.30 ppm. The average pH value of rainwater is 5.6. (The term pH is defined as the negative logarithm of the hydrogen ion concentration in moles per litre. The pH scale ranges from 0 to 14, with lower numbers indicating increased acidity.) On a global basis, as much as 35 percent of the sodium, 55 percent of the chlorine, 15 percent of the potassium, and 37 percent of the sulfate in river water may be derived from the oceans through sea aerosol generation.

A considerable amount of data has become available for marine aerosols. These aerosols are important because (1) they are vital to any description of the global biogeochemical cycle of an element, (2) they may have an impact on climate, (3) they are a sink, via heterogeneous chemical reactions, for trace atmospheric gases, and (4) they influence precipitation of cloud and rain droplets. For many trace metals, the ratio of the atmospheric flux to the riverine flux for coastal and remote oceanic areas may be greater than one, indicating the importance of atmospheric transport. Figures have been prepared that illustrate the Enrichment Factors (EF) of North Atlantic marine aerosols and suspended matter in North Atlantic waters relative to the crust (that is, terrestrial sources), where

$$EF_{crust} = \frac{(X/A1)_{air}}{(X/A1)_{crust}},$$

and $(X/Al)_{air}$ and $(X/Al)_{crust}$ refer, respectively, to the ratio of the concentration of the element X to that of Al, aluminum (which is an easily observed terrestrial component of aerosols), in the atmosphere and in average crustal material. Comparing the enrichment factors in marine aerosols with those of suspended matter in the water column indicates qualitatively the marine aerosols' importance as a source that alters the composition of marine suspended matter and, consequently, their importance to deep-sea sedimentation. Moreover, such comparisons help identify how significant terrestrial sources are for both the marine aerosols and the water below.

In some instances the ratios of ions in rainwater deviate significantly from those in seawater. Mechanisms proposed for this fractionation are, for example, the escape of chlorine as gaseous hydrogen chloride (HCl) from sea salt aerosol with a consequent enrichment in sodium and bubbling and thermal diffusion. In addition, release of biogenic gases such as Dimethyl Sulfide (DMS) from the sea surface and its subsequent reaction in the oceanic atmosphere to sulfate can change rainwater ion ratios with respect to seawater. Soil particles also can influence rainwater composition. Rainfall over the southwestern United States contains relatively high sulfate concentrations because of sulfate-bearing particles that have been blown into the atmosphere from desert soils. Rain near industrial areas commonly contains high contents of sulfate, nitrate, and carbon dioxide (CO_2) largely derived from the burning of coal and oil. There are two main processes leading to the conversion of sulfur dioxide (SO_2) to sulfuric acid (H_2SO_4). These are reactions with hydroxyl radicals (–OH) and with hydrogen peroxide (H_2O_2) in the atmosphere:

$$SO_2 + OH \rightarrow \text{intermediate species} \rightarrow H_2SO_4$$

and,

$$SO_2 + H_2O_2 = H_2SO_4.$$

The sulfuric acid then dissociates to hydrogen and sulfate ions:

$$H_2SO_4 = 2H^+ + SO_4^{2-}.$$

For the nitrogen gases nitric oxide (NO) and nitrogen dioxide (NO_2) released from fossil fuel burning, their atmospheric reactions lead to the production of nitric acid (HNO_3) and its dissociation to hydrogen ions (H^+) and nitrate (NO_3^-). These reactions are responsible for the acid rain conditions that occurred in the northeastern United States, southeastern Canada, and western Europe during the second half of the 20th century. The high sulfate values of the rain in the northeastern United States reflect the acid precipitation conditions of this region.

River and Ocean Waters

River discharge constitutes the main source for the oceans. Seawater has a more uniform composition than river water. It contains, by weight, about 3.5 percent dissolved salts, whereas river water has only 0.012 percent. The average density of the world's oceans is roughly 2.75 percent greater than that of typical river water. Of the average 35 parts per thousand salts of seawater, sodium and chlorine make up almost 30 parts, and magnesium and sulfate contribute another four parts. Of the remaining one part of the salinity, calcium and potassium constitute 0.4 part each and carbon, as carbonate and bicarbonate, about 0.15 part. Thus, nine elements (hydrogen, oxygen, sulfur, chlorine, sodium, magnesium, calcium, potassium, and carbon) make up 99 percent of seawater, though most of the 94 naturally occurring elements have been detected therein. Of importance are the nutrient elements phosphorus, nitrogen, and silicon, along with such essential micronutrient trace elements as iron, cobalt, and copper. These elements strongly regulate the organic production of the world's oceans.

A portion of the delta of the Mekong River as it flows through southern Vietnam and empties into the South China Sea.

In contrast to ocean water, the average salinity of the world's rivers is low—only about 0.012 percent, or 120 ppm by weight. Of this salt content, carbon as bicarbonate constitutes 58 parts, or 48 percent, and calcium, sulfur as sulfate, and silicon as dissolved monomeric silicic acid make up a total of about 39 parts, or 33 percent. The remaining 19 percent consists predominantly of chlorine, sodium, and magnesium in descending importance. It is obvious that the concentrations and relative proportions of dissolved species in river waters contrast sharply with those of seawater. Thus, even though seawater is derived in part by the chemical differentiation and evaporation of river water, the processes involved affect every element differently, indicating that simple evaporation and concentration are entirely secondary to other processes.

Water-Rock Interactions as Determining River Water Composition

Generally speaking, the composition of river water, and thus that of lakes, is controlled by water-rock interactions. The attack of carbon dioxide-charged rain and soil waters on the individual minerals in continental rocks leads to the production of dissolved constituents for lakes, rivers, and streams. It also gives rise to solid alteration products that make up soils or suspended particles in freshwater aquatic systems. The carbon dioxide content of rain and soil waters is of particular importance in weathering processes. The pH of rainwater equilibrated with the atmospheric carbon dioxide partial pressure of $10^{-3.5}$ (0.00032) atmosphere is 5.6. In industrial regions, rainwater pH values may be lower because of the release and subsequent hydrolysis of acid gases—namely, sulfur dioxide and nitrogen oxides (NO_x) from the combustion of fossil fuels. After rainwater enters soils, its characteristics change markedly. The usual few parts per million of salts in rainwater increase substantially as the water reacts. The upper part of the soil is a zone of intense biochemical activity. The bacterial population near the surface is large, but it decreases rapidly downward. One of the major biochemical processes of the bacteria is the oxidation of organic material, which leads to the release of carbon dioxide. Soil gases obtained above the zone of water saturation may contain 10 to 40 times as much carbon dioxide as the free atmosphere, and in some cases carbon dioxide has been shown to make up 30 percent of the soil gases as opposed to 0.03 percent of the free atmosphere. In addition to the acid effects of carbon dioxide, there is a highly acidic microenvironment created by the roots of living plants. Values of pH as low as 2 have been measured immediately adjacent to root hairs. The combined length of a plant's root hairs may be several kilometres, so their chemical effects on acidic water are formidable.

The Amazon River near Manaus, Brazil. The brown colour of the water is the result of rains washing soil particles into the stream from surrounding land and the stirring up of mud on the riverbed.

Congruent and Incongruent Weathering Reactions

These acid solutions in the soil environment attack the rock minerals, the bases of the system, producing neutralization products of dissolved constituents and solid particles. Two general types of reactions occur: congruent and incongruent. In the former, a solid dissolves, adding elements to the water according to their proportions in the mineral. An example of such a weathering reaction is the solution of calcite ($CaCO_3$) in limestones:

$$CaCo_3 + CO_2(g) + H_2O = Ca^{2+} + 2HCO_3^-.$$

Here one of the HCO_3^- ions comes from calcite and the other from $CO_2(g)$ in the reacting water. The amount of carbon dioxide dissolved according to reaction depends on temperature, pressure, original bicarbonate content of the weathering solution, and partial pressure of the carbon dioxide. The carbon dioxide and the temperature are the most important variables. Increases in one or both of these variables lead to increases in the amount of calcite dissolved. For example, for a carbon dioxide pressure of $10^{-3.5}$ (0.00032) atmosphere, the amount of calcium that can be dissolved until saturation is about $10^{-3.3}$ (0.0005) mole, or 20 ppm, at 25 °C (77 °F). For an atmospheric carbon dioxide pressure of 10^{-2} (0.01) atmosphere and for a soil atmosphere of nearly pure carbon dioxide, the values are 65 and 300 ppm, respectively. The weathering of calcite leads to the release of calcium and bicarbonate ions into soil waters and groundwaters, and these constituents eventually reach lake and river systems. The insoluble residue of quartz (SiO_2), clay minerals, and iron oxides (e.g., FeOOH) in the limestone rock make up the deep-red soils that form from limestone weathering. These particles may be carried into streams by runoff and hence to lakes and the oceans and become part of the suspended load of these systems.

An example of an incongruent weathering reaction—that is, one where only part of a solid is consumed—is that involving aluminosilicates. One such reaction is the aggressive attack of carbon dioxide-charged soil water on the mineral K-spar ($KAlSi_3O_8$), an important phase found in continental rocks. The reaction is:

$$2KAlSi_3O_8 + 2CO_2 + 11H_2O = Al_2Si_2O_5(OH)_4 + 2K^+ + 2HCO_3^- + 4H_4SiO_4.$$

It should be noted that the K-spar changes into a new mineral—kaolinite (a clay mineral) in this case—plus solution, and acid is consumed. The total dissolved material per litre of soil solution released is about 60 ppm for a solution initially equilibrated with a typical soil carbon dioxide content. The water resulting from reaction would contain bicarbonate, potassium, and dissolved silica in the ratios 1:1:2, and the new solid, kaolinite, would be a weathering product. These dissolved constituents and the solid alteration product would eventually reach rivers to be transferred possibly to lakes and ultimately to the sea. It has been demonstrated that the composition of river water is the product of a variety of mineral-water reactions such as. The dissolved load of the world's rivers comes from the following sources: 7 percent from beds of halite (NaCl) and salt disseminated in rocks, 10 percent from gypsum ($CaSO_4 \cdot 2H_2O$) and anhydrite ($CaSO_4$) deposits and sulfate salts disseminated in rocks, 38 percent from limestones and dolomites, and 45 percent from the weathering of one silicate mineral to another. Of the bicarbonate ions in river water, 56 percent stems from the atmosphere, 35 percent from carbonate minerals, and 9 percent from the oxidative weathering of fossil organic matter. Reactions involving silicate minerals account for 30 percent of the riverine bicarbonate ions.

Besides dissolved substances, rivers also transport solids in traction (i.e., bed load) and, most importantly, suspended load. The present global river-borne flux of solids to the oceans is estimated as 15.5 billion metric tons (about 17.1 billion tons) per year. Present elemental fluxes are estimated in millions of metric tons per year as silicon, 4,420; aluminum, 1,460; iron, 740; calcium, 330; potassium, 310; magnesium, 210;

and sodium, 110. The total load of particulate organic carbon of the world's rivers is 180 million metric tons (nearly 200 million tons) per year. The riverine fluxes of trace metals to the oceans are dominated by their occurrence in the particulate phase as opposed to the dissolved phase. The particulate matter in river water is an important source of silicon, aluminum, iron, titanium, rubidium, scandium, vanadium, the lanthanoids, and other elements for deep-sea sediments.

Lake Waters

Although lake waters constitute only a small percentage of the water in the hydrosphere, they are an important ephemeral storage reservoir for fresh water. Aside from their recreational use, lakes constitute a source of water for household, agricultural, and industrial uses. Lake waters are also very susceptible to changes in chemical composition due to these uses and to other factors.

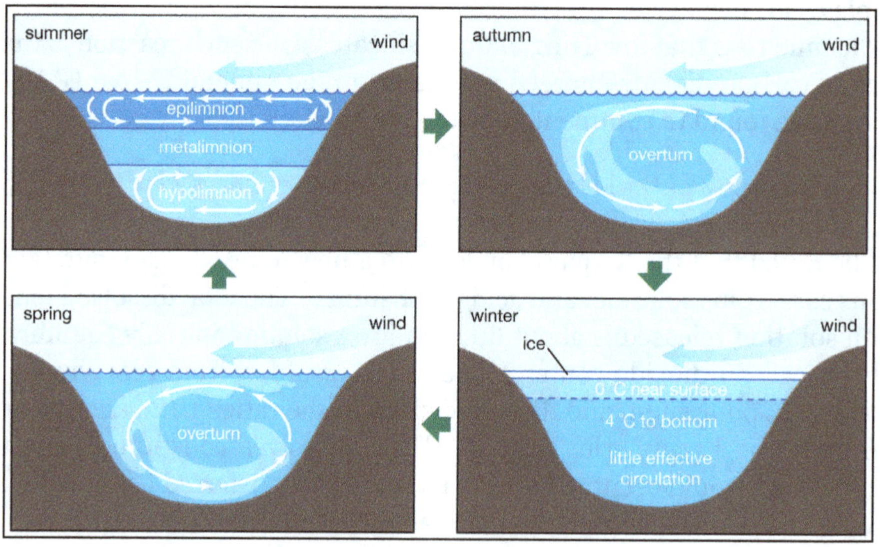

Annual circulation patterns in a dimictic lake. The typical dimictic lake has distinct layers that fully mix twice a year. It undergoes stratification in the summer and complete overturn in the autumn and spring. During winter, surface ice prevents further mixing by the wind. Small differences in density and temperature exist, with cooler water (0 °C [32 °F]) staying near the surface and warmer, denser water (4 °C [39.2 °F]) extending to the bottom.

In general, fresh waters at the continental surface evolve from their rock sources by enrichment in calcium and sodium and by depletion in magnesium and potassium. In very soft waters the alkalies may be more abundant than the alkaline earths, and in the more-concentrated waters of open river systems calcium > magnesium > sodium > potassium. For the anions, in general, HCO_3^- exceeds SO_4^{2-}, which is greater in concentration than Cl^-. It is worthwhile at this stage to consider some major mechanisms that control global surface water composition. These mechanisms are atmospheric precipitation, rock reactions, and evaporation-precipitation.

The mechanism principally responsible for waters of very low salinity is precipitation. These waters tend to form in tropical regions of low relief and thoroughly leached source rocks. In these regions rainfall is high, and volumes of fresh water (rivers, tributary streams, pools, etc.) within a watershed are usually dominated by salts brought in by precipitation. Such waters constitute one part of a chain of water volumes that begins with falling precipitation and ends with the release of water into the ocean, for which the final part of the chain represents water volumes dominated by contributions of dissolved salts from the rocks and soils of their basins. These waters have moderate salinity and are rich in dissolved calcium and bicarbonate. They are in turn the "end-member" of another series that extends from the calcium-rich, medium-salinity fresh waters to the high-salinity, sodium chloride-dominated waters of which seawater is an example. Seawater composition, however, does not evolve directly from the composition of fresh waters and the precipitation of calcium carbonate; other mechanisms that control its composition are involved. Such factors as relief and vegetation also may affect the composition of the world's surface waters, but atmospheric precipitation, water-rock reactions, and evaporation-crystallization processes appear to be the dominant mechanisms governing continental surface water chemistry.

Continental fresh waters evaporate once they have entered closed basins, and their constituent salts precipitate on the basin floors. The composition of these waters may evolve along several different paths, depending on their initial chemical makeup.

Biological processes strongly affect the composition of lake waters and are responsible to a significant degree for the compositional differences between the upper water layer (the epilimnion) and the lower water layer (the hypolimnion) of lakes. The starting point is photosynthesis, represented by the following reaction:

$$106\,CO_2 + 16\,NO_3^- + HPO_4^{2-} + 122\,H_2O + 18\,H^+ \xrightarrow[\text{micro nutrients}]{\text{radiant energy}} C_{106}H_{263}O_{110}N_{16}P + 138\,O_2\,.$$

The reversal of this reaction is oxidation-respiration leading to the release of the nutrients nitrogen and phosphorus, as well as carbon dioxide. In a stratified lake, carbon, nutrients, and silica are extracted from the upper layer during photosynthesis. This process leads to reduced concentrations of nitrate, phosphate, and silica in these waters and, during times of maximum daylight organic production, to supersaturation of the upper layer with respect to dissolved oxygen. The organic matter produced by phytoplankton may be either grazed upon by zooplankton and other organisms or decomposed by bacteria. Some of it, however, sinks into the lower layer. There it is further decomposed, especially by bacteria, resulting in the release of dissolved phosphorus and nitrogen and the consumption of oxygen. Oxygen concentrations therefore are reduced in these lower lake waters, because stratification prevents oxygen exchange with the atmosphere. Furthermore, the inorganic carbonate and siliceous skeletons of the dead organisms sinking into the lower layer may dissolve, giving rise to increased concentrations of dissolved silica and inorganic carbon in the deep waters of stratified lakes. This dissolution is a result of undersaturation of the waters of the lower layer with respect to the opaline silica and calcium carbonate that make up the skeletons of the dead and sinking plankton. These natural biological processes have been accelerated in some lakes because of excess nutrient input by human activity, resulting in the eutrophication of lake waters and marine systems.

Groundwaters

Groundwaters derive their compositions from a variety of processes, including dissolution, hydrolysis, and precipitation reactions; adsorption and ion exchange; oxidation and reduction; gas exchange between groundwater and the atmosphere; and biological processes.

Processes affecting the major chemical components of groundwater	
Component	Origin
Sodium ion	Sodium chloride dissolution (some pollutive)
	Plagioclase weathering
	Rainwater addition
Potassium ion	Biotite weathering
	K-feldspar weathering
	Biomass decreases
	Dissolution of trapped aerosols
Magnesium ion	Amphibole and pyroxene weathering
	Biotite (and chlorite) weathering
	Dolomite weathering
	Olivine weathering
	Rainwater addition
Calcium ion	Calcite weathering
	Plagioclase weathering
	Dolomite weathering
	Dissolution of trapped aerosols
	Biomass decreases
Bicarbonate ion	Calcite and dolomite weathering
	Silicate weathering
Sulfate ion	Pyrite weathering (some pollutive)
	Rainwater addition
Chloride ion	Sodium chloride dissolution (some pollutive)
	Rainwater addition
Hydrogen silicate	Silicate weathering

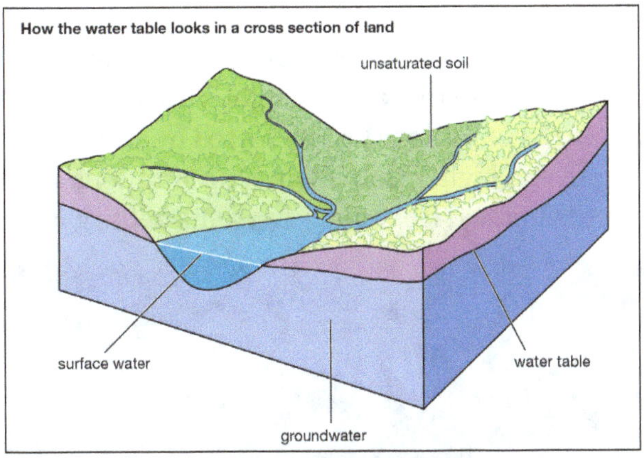

The water table is the top level of groundwater. Surface water is an exposed part of the water table.

The biological processes of greatest importance are microbial metabolism, organic production, and respiration (oxidation). By far the most important overall process for the major constituents of groundwater is that of mineral-water reactions. Thus, the composition of groundwaters strongly reflects the types of rock minerals that the waters have encountered in their movement through the subsurface.

In general, the most mobile elements in groundwater—i.e., those most easily liberated by the weathering of rock minerals—are calcium, sodium, and magnesium. Silicon and potassium have intermediate mobilities, and aluminum and iron are essentially immobile and locked up in solid phases.

Groundwaters are highly susceptible to contamination because of human activities and the fact that their dissolved constituents are derived to a large extent from the leaching of surface materials. Some of the nitrogen and phosphorus applied to soils as fertilizers and organic pesticides may be leached and leak into groundwater systems, leading to increased concentrations of ammonium and phosphate. Radioactive wastes, industrial chemicals, household materials, and mine refuse are other anthropogenic sources of dissolved substances that have been detected in groundwater systems.

Ice

Ice is nearly a pure solid and, as such, accommodates few foreign ions in its structure. It does contain, however, particulate matter and gases, which are trapped in bubbles within the ice. The change in composition of these materials through time, as recorded in the successive layers of ice, has been used to interpret the history of Earth's surface environment and the impact of human activities on this environment. The increase in the lead content of continental glacial ice with decreasing age of the ice up to the middle of the 1970s, for example, reflects the progressive input of tetraethyl lead into the global environment from gasoline burning. (Stringent environmental regulations that appeared in the 1970s regarding the use of leaded gasoline has led to a fall in lead concentrations in ice laid down since that time.) Also, atmospheric carbon dioxide and methane concentrations, which have increased significantly during the past century because of anthropogenic activities, are faithfully recorded in ice bubbles of the thick continental ice sheets. By 2016 atmospheric carbon dioxide and methane concentrations had increased by more than 43 percent and more than 150 percent, respectively, higher than their concentrations 200 years ago; the latter concentration values were obtained from measurements of the gases in air trapped in ice.

Perito Moreno Glacier, Los Glaciares National Park.

The Water Cycle

General Nature of the Cycle

The present-day water cycle at Earth's surface is made up of several parts. Some 496,000 cubic km (about 119,000 cubic miles) of water evaporates from the land and ocean surface annually, remaining for about 10 days in the atmosphere before falling as rain or snow. The amount of solar radiation necessary to evaporate this water is half of the total solar radiation received at Earth's surface. About one-third of the precipitation falling on land runs off to the oceans primarily in rivers, while direct groundwater discharge to the oceans accounts for only about 0.6 percent of the total discharge. A small amount of precipitation is temporarily stored in the waters of rivers and lakes. The remaining precipitation over land, 73,000 cubic km (17,500 cubic miles) per year, returns to the atmosphere by evaporation. Over the oceans, evaporation exceeds precipitation, and the net difference represents transport of water vapour over land, where it precipitates as rain or snow and returns to the oceans as river runoff and direct groundwater discharge.

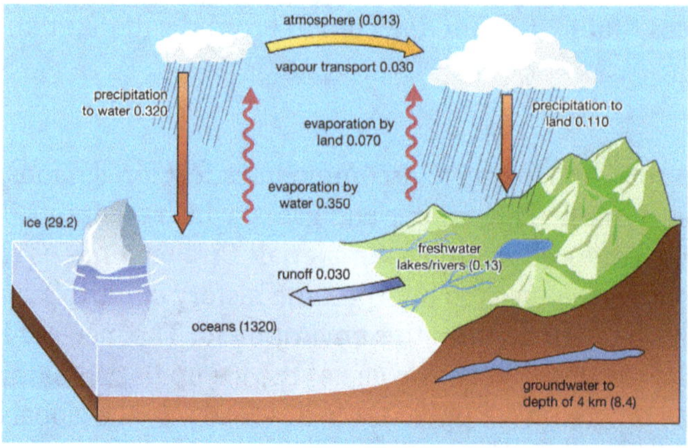

The present-day surface hydrologic cycle, in which water is transferred from the oceans through the atmosphere to the continents and back to the oceans over and beneath the land surface. The values in parentheses following the various forms of water (e.g., ice) refer to volumes in millions of cubic kilometres; those following the processes (e.g., precipitation) refer to their fluxes in millions of cubic kilometres of water per year.

The various reservoirs in the water cycle have different water residence times. Residence time is defined as the amount of water in a reservoir divided by either the rate of addition of water to the reservoir or the rate of loss from it. The oceans have a water residence time of 3,000 to 3,230 years; this long residence time reflects the large amount of water in the oceans. In the atmosphere the residence time of water vapour relative to total evaporation is only about 10 days. Lakes, rivers, ice, and groundwaters have residence times lying between these two extremes and are highly variable.

There is considerable variation in evaporation and precipitation over the globe. In order for precipitation to occur, there must be sufficient atmospheric water vapour and enough rising air to carry the vapour to an altitude where it can condense and precipitate. Precipitation and evaporation vary with latitude and their relation to the global wind belts. The trade winds, for example, are initially cool, but they warm up as they blow toward the Equator. These winds pick up moisture from the ocean, increasing ocean surface salinity and causing seawater at the surface to sink. When the trade winds reach the Equator, they rise, and the water vapour in them condenses and forms clouds. Net precipitation is high near the Equator and also in the belts of the prevailing westerlies, where there is frequent storm activity. Evaporation exceeds precipitation in the subtropics, where the air is stable, and near the poles, where the air is both stable and has a low water vapour content because of the cold. The Greenland Ice Sheet and the Antarctic Ice Sheet formed because the very low evaporation rates at the poles resulted in precipitation exceeding evaporation in these local regions.

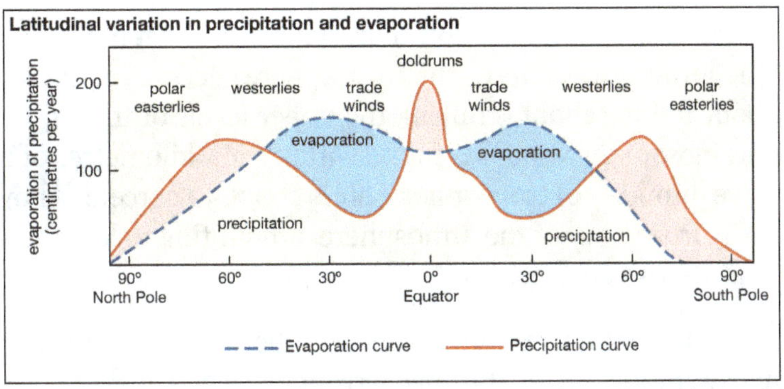

Latitudinal variation in precipitation and evaporation and its relationship to major wind belts and oceanic salinity.

Processes Involved in the Cycle

The water cycle consists of various complicated processes that move water throughout the different reservoirs on the planet. The major processes involved are precipitation, evaporation, interception, transpiration, infiltration, percolation, retention, detention, overland flow, throughflow, and runoff.

Water Vapour and Precipitation

Water exists in the atmosphere in gaseous form. Its liquid form, either as water droplets in clouds or as rain, and its solid form, as ice crystals in clouds, snowflakes, or hail, occur only momentarily and locally.

Water vapour performs two major functions: (1) it is important to the radiation balance of Earth, as its presence keeps the planetary surface warmer than would otherwise be the case, and (2) it is the principal phase of the ascending part of the water cycle.

The mass of water vapour in the atmosphere, which represents only 0.001 percent of the hydrosphere, is highest in the tropics and decreases toward the poles. At a mean temperature of Earth's surface of 15 °C (59 °F), the partial pressure of water vapour at equilibrium with pure water is 0.017 atmosphere. The addition of salts to pure water lowers its vapour pressure. The equilibrium, or saturation, water vapour pressure of a saturated solution of sodium chloride is 22 percent lower than that of pure water. Precipitable water vapour has, on the average, a vapour pressure of 0.0025 atmosphere, which amounts to 15 percent of the saturation vapour pressure. The ratio of observed water vapour pressure to the saturation vapour pressure at the same temperature is the relative humidity of the air. Thus, the mean relative humidity of the atmosphere is only 15 percent, a value that is low in human terms, in which levels of 50 to 60 percent are preferred for maximum comfort. The relative humidity of the air, however, varies greatly from one geographic region to another and also vertically in the atmosphere. Atmospheric water vapour decreases rapidly with increasing altitude relative to its surface value. The amount of water required to saturate a volume of air depends on the temperature of the air. Air at high temperature can hold more water vapour at saturation than can air at low temperature. Because the temperature of the lower atmosphere (the troposphere) decreases rapidly with increasing altitude to about 15 km (about 9 miles), the upper levels of the troposphere contain little water vapour; most of the vapour is found within a few kilometres of Earth's surface. The average relative humidity of tropospheric air is about 50 percent. Above 15 km, water vapour is essentially frozen out of the atmosphere, amounting to less than 0.1 percent of its concentration at Earth's surface.

Aside from temperature, other factors determine the water vapour content of the air and are particularly important in the lower troposphere. These factors include local evaporation and the horizontal atmospheric transportation of moisture, which varies with altitude, latitude, season, and topography. During a period of 10 days (i.e., the residence time of water in the atmosphere), horizontal eddy turbulence may disperse vapour over distances up to 1,000 km (600 miles).

When a mass of air at Earth's surface is exposed to a body of water, it gains water by evaporation or loses water by precipitation, depending on its relative humidity. If the air is undersaturated, with a relative humidity of less than 100 percent, it gains water vapour because the rate of evaporation exceeds the rate of condensation. If the air is supersaturated, with a relative humidity greater than 100 percent, the air mass loses water vapour because the rate of precipitation exceeds that of evaporation. This interaction between air masses and surface water bodies drives the atmosphere toward a state of saturation, which is not achieved for the entire atmosphere because of the variability in weather and because not all air masses are in contact with water bodies. In general, the level of atmospheric water vapour is higher in the summer, since temperatures are higher at this time of year. Also, atmospheric water vapour content is higher near the source of moisture than in distant regions. Over the oceans, the air is almost always near saturation, whereas over the deserts, where the supply of moisture is limited, the air is far below water vapour saturation values. In most cases, atmospheric water vapour content decreases inland over continents, but this decrease is modified by rainfall conditions, by the presence or absence of high mountains, large lakes, extensive forests, and swamps, and by the prevailing wind directions. Horizontal winds and air mass movements transfer water vapour from the ocean to the land. Although the processes are not completely separable, the horizontal transfer of water vapour seldom causes the vapour to undergo condensation, whereas vertical movements are most important in the condensation process.

Condensation depends strongly on the average temperature of Earth's surface because the water vapour content of the air is strongly dependent on temperature. In figures that show the states of water as a function of the variables of pressure and temperature, the slope of the phase boundary between liquid water and water vapour is positive, implying that with increasing temperature the air at equilibrium will hold increasing amounts of water vapour. Cooling or mixing of this air results in condensation of the vapour and precipitation as water droplets or as ice crystals if the air temperature is below 0 °C (32 °F). When first formed, the water droplets or ice crystals are very small, on the order of 10^{-2} to 10^{-3} cm (0.004 to 0.0004 inch) in diameter, and they float freely in the atmosphere. In large quantities, these water droplets and ice crystals produce a cloud. All clouds are formed as a result of cooling below the dew point, the temperature at which condensation begins when air is cooled at constant pressure and constant water vapour content. When the droplets or crystals coalesce to a size of about 10^{-2} cm (0.004 inch) in diameter, they become heavy enough to fall as raindrops or snowflakes. Hailstones measure about 10^{-1} cm (0.04 inch) in diameter or much larger. Water vapour condensing in the atmosphere contains strongly soluble salts (mostly of oceanic origin), weakly soluble or insoluble solids (dust), and dissolved gases. The dust and sea salt aerosol particles in the air may act as sites of condensation by serving as nuclei for bringing initially a few water molecules together and inducing condensation from supersaturated air.

Distribution of Precipitation

Precipitation falling toward Earth's surface may suffer several fates. It may be evaporated during its fall or after it reaches the ground surface. If the surface is covered with dense vegetation, much of the precipitation may be held on leaves and plant limbs and stems. This process is termed interception and may result in little water reaching the ground because the water may be directly evaporated from plant surfaces back into the atmosphere. If precipitation reaches the ground in the form of snow, it may remain there for some time. On the other hand, if precipitation falls as rain, it may evaporate, infiltrate the soil, be detained in small catchment areas, or become overland flow—a form of runoff. Overland flow (R_o) may be expressed in terms of intensity units, water depth per unit of time (e.g., centimetres per hour, or inches per hour) as,

$$R_o = P - I,$$

where P is precipitation rate and I is infiltration rate (rate of entry and downward movement of water into the soil profile). Infiltration rate will equal precipitation rate until the limit of the infiltration rate, or infiltration capacity, is reached. Soil infiltration rates are usually high at the beginning of a rain preceded by a dry spell and decrease as the rainfall continues. This change in rate is due to the clogging of soil pores by particles brought from above by the infiltrating rain and to the swelling of colloidal soil particles as they absorb water. Thus, rapid decreases in infiltration rates during a rain are more likely to occur in clay-rich soils than in sandy soils.

Between rainfall periods, water held in the soil as moisture is gradually lost by direct evaporation or by withdrawal by plants. Evaporation into the open atmosphere occurs at the surface of the soil, and the soil dries progressively downward with time. Water vapour in the soil diffuses upward, replenishing the evaporated water, and in turn is evaporated. The pumping of air into and out of the soil by atmospheric pressure changes enhances the movement of soil moisture upward. It has been shown that evaporation of a water droplet in the free atmosphere, and to a first approximation in various soil atmospheres, is proportional to the droplet surface area $4\pi r^2$ (square centimetres, where r is the radius of the droplet), the diffusional flux of water at the droplet surface, and the transfer of heat as the droplet evaporates. The equation for the rate of shrinkage of a water droplet due to evaporation is,

$$\frac{dr}{dt} = \frac{D_{\rho vo}(Sp - 1)}{r_{\rho L}(1 + X)}$$

where dr/dt is the rate of change in the radius of the water droplet (centimetres per second), D is the diffusion coefficient of water vapour in air (cubic centimetres per second), P_{vo} is the equilibrium vapour concentration at the droplet surface, Sp is the degree of undersaturation of water vapour in the environment, ρ_L is the density of liquid water (grams per cubic centimetre), and X is a dimensionless parameter depending on D, ρ_{vo}, temperature, the heat of evaporation of water vapour, the coefficient of thermal conductivity of air, and the spherical coordinate system necessary to define processes occurring to a spherical water droplet. Water droplets shrink—dr/dt < 0, evaporate—when the water vapour concentration in the environment (atmosphere or soil atmosphere) is less than the saturation water vapour concentration at the droplet surface. They grow—dr/dt > 0, condense—when the converse is true in the free atmosphere. The term dr/dt has negative values for evaporation and positive ones for condensation. Use of this equation shows, as an example, that it would take 23 minutes for a water droplet to shrink (evaporate) in size from 50 to 5 micrometres (0.002 to 0.0002 inch) in air at 10 °C (50 °F) and a water vapour undersaturation of 1 percent.

Besides simple evaporation of water from soils, water is also returned to the atmosphere by transpiration in plants. Plants draw water from soil moisture through their vast network of root hairs and rootlets. This water is carried upward through the plant trunk and branches into the leaves, where it is discharged as water vapour. The term evapotranspiration is used in climatic and hydrologic studies to include the combined water loss from Earth's surface resulting from evaporation and transpiration. The maximum possible evapotranspiration is termed potential evapotranspiration and is governed by the available heat energy. It is taken as equal to evaporation from a large water surface and is generally much less than actual evapotranspiration. Actual evapotranspiration is never greater than precipitation except on irrigated land because of percolation of water into groundwater bodies and surface runoff.

The soil moisture zone gains water by precipitation and infiltration and loses water by evapotranspiration, overland flow, and percolation of water downward due to gravity into the groundwater zone. The contact between the groundwater zone (phreatic zone) and the overlying unsaturated zone (vadose zone) is called the groundwater table. The water balance equation for change of moisture storage in a soil is given as:

$$S = P - E - R,$$

where S is storage, P is precipitation, E is evaporation, and R is surface runoff plus percolation rate into the groundwater zone; all terms are in units of length per unit of time (e.g., millimetres per day, centimetres per month). In humid midlatitude climates where a strong contrast between winter and summer temperatures exists, there is an annual cycle of the water content of soils. The annual cycle of moisture in soil in Ohio, U.S., for example, demonstrates the processes controlling soil moisture. Of special importance is the fact that the soils are saturated in this temperate climate in the spring, and the evaporation rate is low because of the low input of radiant energy from the Sun. By contrast, in the summer, evaporation increases because of increasing solar radiation, and with the growth of plants so does transpiration. Soil moisture is reduced to very low levels at this time of year.

Groundwaters and River Runoff

The term R in the water balance equation for change of soil moisture storage above represents groundwater and river runoff losses from the soil moisture zone. Water percolates from the soil moisture zone through the unsaturated (vadose) zone to the water table. Flow through the unsaturated zone is complicated. After a rainfall, water may form a nearly continuous phase in pores in this zone, but, with drying, the last amount of water is held in clusters at points of contact of solid grains and as thin films on solid surfaces. The flow paths of water become more tortuous, and the water-conducting properties decrease rapidly. Structured soils and fractured rock in the vadose zone may act as conduits for fluids to reach the water table. Because of the complex geometry of water contained in the unsaturated zone, the properties of water are expressed by means of empirical relationships. Darcy's law, derived in 1856 from experimentation by the French engineer Henri Darcy, permits quantification of water flow through porous media. The law states that the rate of flow Q of a fluid through a porous layer of medium (e.g., a sand bed) is directly proportional to the area A of the layer and to the difference Δh between the fluid heads at the inlet and outlet faces of the layer and is inversely proportional to the thickness L of the layer. Expressed analytically,

$$Q = \frac{KA\Delta h}{L},$$

where K is a constant characteristic of the medium. The term K for a porous rock medium is the volume of fluid of unit viscosity passing through a unit cross section of the rock in unit time under the action of a unit pressure gradient; this characteristic is called permeability. The permeability of a rock is dependent on the geometric properties of the rock, such as porosity, shape and size distribution of constituent rock grains, and degree of cementation of the rock. Permeabilities of rocks vary greatly. Unconsolidated sands may have permeabilities measured in hundreds of darcys, whereas consolidated sands that will transmit reasonable amounts of fluid have permeabilities of 0.01 to 1 darcy. A rough idea of the meaning of one darcy of permeability (which equals 9.869×10^{-12} square metre [1.261×10^{-11} square foot]) can be obtained by imagining a cube of sand one foot on a side. If the sand has a permeability of one darcy, approximately one barrel of water per day will pass through the one-foot cube with a one-pound pressure head. The general equation of Darcy can be modified to express flow in both the unsaturated zone and the saturated groundwater zone.

Groundwater is constantly in motion. When a lake or stream intersects the groundwater table, groundwater communicates directly with these bodies of water. If the groundwater table is higher than the stream or lake level, a pressure head will develop such that the groundwater flows into the water body; conversely, if the groundwater table is lower than the river or lake level, the pressure gradient induces flow into the groundwater. Most groundwater ultimately reaches the channels of surface streams and rivers and flows to the sea. On the average, groundwater contributes to total river runoff about 30 percent of its water on a global basis.

Water runoff from the land surface is that part of precipitation which eventually appears in perennial or intermittent surface streams. Streamflow-generation mechanisms have been studied for several decades, and there is now considerable knowledge regarding rainfall runoff processes and their controls. This understanding is the result of both careful observations from field experiments and the heuristic simulations of hypothetical realities with rigorous mathematical models. The discharge measured at the downstream end of a channel reach is supplied by channel inflow at the upstream end of the reach and by the lateral inflows that enter the channel from the hillslope along the reach. The lateral inflows may arrive at the stream in one of three forms: (1) groundwater flow, (2) subsurface storm flow, or (3) overland flow.

Groundwater flow provides the base flow component of streams that sustains their flow between storms. The "flashy" response of streamflow to individual precipitation events may be ascribed to either subsurface storm flow or overland flow. Under intense rainfall events during which the surface soil layer becomes saturated to some depth, water is able to migrate through "preferred pathways" rapidly enough to deliver contributions to the stream during the peak runoff period. The conditions for subsurface storm flow are quite restrictive. The mechanism is most likely to be operative on steep, humid, forested hillslopes with very permeable surface soils.

Overland flow is generated at a point on a hillslope only after surface ponding takes place. Ponding cannot occur until the surface soil layers become saturated. It is now widely recognized that surface saturation can occur because of two quite distinct mechanisms—specifically, Horton overland flow (named for American hydraulic engineer and hydrologist Robert E. Horton) and Dunne overland flow (named for British hydrologist Thomas Dunne).

The former classic mechanism is for a precipitation rate that exceeds the saturated hydraulic conductivity of the surface soil. A moisture content versus depth profile during such a rainfall event will show moisture contents that increase at the surface as a function of time. At some point in time the surface becomes saturated, and an inverted zone of saturation begins to propagate downward into the soil. It is at this time that the infiltration rate drops below the rainfall rate and overland flow is generated. The time is called the ponding time. The necessary conditions for the generation of overland flow by the Horton mechanism are (1) a rainfall rate greater than the saturated hydraulic conductivity of the soil and (2) a rainfall duration longer than the required ponding time for a given initial moisture profile. Horton overland flow is generated from partial areas of the hillslope where surface hydraulic conductivities are lowest.

In Dunne overland flow, the precipitation rate is less than the saturated hydraulic conductivity, and the initial water table is shallow or there is a shallow impeding layer. Surface saturation occurs because of a rising water table; ponding and overland flow occur at a time when no further soil moisture storage is available. The Dunne mechanism is more common to near-channel areas. Dunne overland flow is generated from partial areas of the hillslope where water tables are shallowest. Both Horton and Dunne mechanisms result in variable source areas that expand and contract through wet and dry periods.

Total river discharge and the chemistry of the discharge vary from continent to continent; some continents are wetter and some drier than the world average, but the deviations are not extreme. The runoff per unit area from Asia and Europe is almost exactly equal to the world average; it is a little lower in Africa and North America; and it is considerably higher in South America. Antarctica is frozen and Australia is arid, and so they contribute little runoff. Also, since their areas are relatively small, they do not affect the global runoff average significantly. The waters draining the continents have quite different chemistries; those from Europe are very rich in calcium and bicarbonates, whereas those from Africa and South America are not. North American and Asian rivers are somewhat intermediate in their concentrations of these dissolved constituents. Such differences in composition reflect a variety of factors, including runoff, temperature, and relief, but certainly the bulk composition of the continental rocks in contact with these waters and their underground sources play a major role. The surface rocks of Europe are rich in carbonates, and those of South America are not; the latter are dominated by sediments rich in silicate minerals.

The chemistry of groundwater and river runoff is being modified by human activities on a global scale. The natural dissolved riverine input of major constituents to the oceans already has been increased by more than 10 percent because of human activities. In the case of sodium, chlorine, and sulfate, the increases are as high as 30 percent. In the United States alone, total water use is equivalent to one-third of total runoff, with about 2 percent of the water used coming from underground wells. In the southwestern region of the country, water supplies have been tapped heavily and in some areas have been exhausted with no hope of replacement. This extensive use of fresh waters in the United States and throughout the globe makes them particularly susceptible to pollution. Leachates from fertilizers, herbicides, and pesticides are found in some freshwater bodies; toxins or excessive amounts of certain inorganic or organic chemicals are present; radioactive elements have been detected; and some surface water bodies have had their salinities increased dramatically, rendering them useless for human consumption. It is therefore imperative that countries closely monitor the use of freshwater systems and promote their conservation.

The Early Hydrosphere

The gases released from Earth during its early history, including water vapour, have been called excess volatiles because their masses cannot be accounted for simply by rock weathering. These volatiles are thought to have formed the early atmosphere of Earth. At an initial crustal temperature of about 600 °C (about 1,100 °F), almost all of these compounds, including water (H_2O), would have been in the atmosphere. The sequence of events that occurred as the crust cooled is difficult to reconstruct. Below 100 °C (212

°F) all of the water would have condensed, and the acid gases would have reacted with the original igneous crustal minerals to form sediments and an initial hydrosphere that was dominated by a salty ocean. If the chemical reaction rates are assumed to have been slow relative to cooling, an atmosphere of 600 °C would have contained, together with other compounds, water vapour, carbon dioxide, and hydrogen chloride (HCl) in a ratio of 20:3:1 and cooled to the critical temperature of water (that is, the temperature above which a compound cannot remain as a liquid—i.e., 374 °C [705 °F]). The water therefore would have condensed into an early hot ocean. At this stage, the hydrogen chloride would have dissolved in the ocean (about one mole per litre), but most of the carbon dioxide would have remained in the atmosphere, with only about 0.5 mole per litre in the ocean water. This early acid ocean would have reacted vigorously with crustal minerals, dissolving out silica and cations and creating a residue composed principally of aluminous clay minerals that would form the sediments of the early ocean basins.

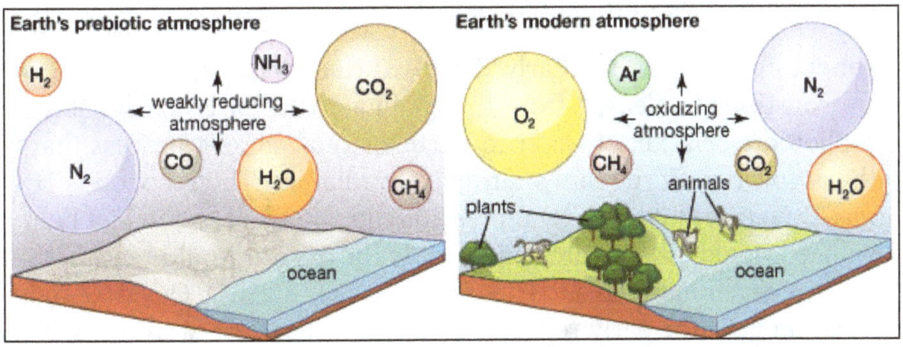

Earth's early and modern atmospheres.

This is one of several possible pathways for the early surface of Earth. Whatever the actual case, after Earth's surface had cooled to 100 °C, it would have taken only a short time for the remaining acid gases to be consumed in reactions involving igneous rock minerals. The presence of cyanobacteria (e.g., blue-green algae) in the fossil record of rocks older than three billion years attests to the fact that Earth's surface had cooled to temperatures lower than 100 °C by this time, and neutralization of the original acid volatiles had taken place (acid-base reaction). It is possible, however, that, because of increased greenhouse gas concentrations early in the Archean Eon (4.0 billion to 2.5 billion years ago), Earth's surface could still have been warmer than today.

If most of the degassing of primary volatile substances from Earth's interior occurred early, the chloride released by the reaction of hydrochloric acid with rock minerals would be found in the oceans or in evaporite deposits, and the oceans would have a salinity and volume comparable to that of today. This conclusion is based on the assumption that there has been no drastic change in the ratios of volatiles released through geologic time. The overall generalized reaction indicative of the chemistry leading to the formation of the early oceans can be written in the form: primary igneous rock minerals + acid volatiles $^{+}H_2O \rightarrow$ sedimentary rocks + oceans + atmosphere. It should be noted from this equation that, if all the acid volatiles and H_2O were released early in the history of Earth and in the proportions found today, then the total original sedimentary rock mass-produced would be equal to that of the present, and ocean salinity and volume would be close to those of today as well. If, on the other hand, degassing were linear with time, then the sedimentary rock mass would have accumulated at a linear rate, as would have oceanic volume. The salinity of the oceans, however, would remain nearly the same if the ratios of volatiles degassed did not change with time. The most likely situation is the one presented here—namely, that major degassing occurred early in Earth's history, after which minor amounts of volatiles were released episodically or continuously for the remainder of geologic time. The salt content of the oceans based on the constant proportions of volatiles released would depend primarily on the ratio of sodium chloride locked up in evaporites to that dissolved in the oceans. If all the sodium chloride in evaporites were added to the oceans today, the salinity would be approximately doubled. This value gives a sense of the maximum salinity that the oceans could have attained throughout geologic time.

One component absent from the early Earth's surface was free oxygen; it would not have been a constituent released from the cooling crust. Early production of oxygen was by the photodissociation of water in Earth's atmosphere, a process that was triggered by the absorption of the Sun's ultraviolet radiation. The reaction is $H_2O_{(liquid)} + h\nu \rightarrow H_2\,(gas) + O_2\,(gas)$ in which hv represents the photon of ultraviolet light. The hydrogen produced would escape into space, while the oxygen would react with the early reduced gases by reactions such as $2H_2S + 3O_2 \rightarrow 2SO_2 + 2H_2O$. Oxygen production by photodissociation gave the early reduced atmosphere a start toward present-day conditions, but it was not until the appearance of photosynthetic organisms approximately three billion years ago that oxygen could accumulate in Earth's atmosphere at a rate sufficient to give rise to today's oxygenated environment. The photosynthetic reaction leading to oxygen production is given in equation:

$$106\,CO_2 + 16\,NO_3^- + HPO_4^{2-} + 122\,H_2O + 18\,H^+ \xrightarrow[\text{micro nutrients}]{\text{radiant energy}} C_{106}H_{263}O_{110}N_{16}P + 138\,O_2\,.$$

The Transitional Hydrosphere

The nature of the rock record from the time of the first sedimentary rocks (approximately 3.8 billion years ago) to about one to two billion years ago suggests that the amount of oxygen in Earth's atmosphere was significantly lower than it is today and that there were continuous chemical trends in the sedimentary rocks formed and, more subtly, in the composition of the hydrosphere. The chemistry of rocks shifted dramatically during this transitional period. The source rocks of sediments during this time may have been more basaltic than subsequent ones. Sedimentary debris was formed by the alteration of such source rocks in an oxygen-deficient atmosphere and accumulated primarily under anaerobic (oxygen-free) marine conditions. The chief difference between reactions involving mineral-ocean equilibria at this time and the present day was the role played by ferrous iron (i.e., reduced state of iron). The concentration of dissolved iron in modern oceans is low because of the insolubility of oxidized iron oxides. During the transition stage and earlier, oxygen-deficient environments were prevalent, and these favoured the formation of minerals containing ferrous iron from the alteration of rocks slightly more rich in basalt than those of today. Indeed, iron carbonate siderite and iron silicate greenalite, in close association with chert and iron sulfide pyrite, are characteristic minerals that occur in iron formations of the middle Precambrian (about 2.5 billion to 1.6 billion years ago). The chert originally was deposited as amorphous silica; equilibrium between amorphous silica, siderite, and greenalite at 25 °C (77 °F) and a total pressure of one atmosphere requires a carbon dioxide pressure of about $10^{-2.5}$ (0.00316) atmosphere, or 10 times the present-day value.

A banded-iron formation (BIF) rock recovered from the Temagami greenstone belt in Ontario, Canada, and dated to 2.7 billion years ago. Dark layers of iron oxide are intercalated with red chert.

The oceans of this transitional period can be thought of as a solution that resulted from an acid leach of basaltic rocks, and, because the neutralization of the volatile acid gases was not restricted primarily to land areas as it is today, much of this alteration may have occurred by submarine processes. Anaerobic depositional environments with internal carbon dioxide pressures of about $10^{-2.5}$ atmosphere prevailed, and the oxygen-deficient atmosphere itself may have had a carbon dioxide pressure close to $10^{-2.5}$ atmosphere. If so, the pH of early ocean water was lower than that of modern seawater and the calcium concentration was higher; moreover, the early ocean water was probably saturated with respect to amorphous silica—roughly 120 parts per million (ppm).

To simulate what might have occurred, it is helpful to imagine emptying the Pacific basin, throwing in great masses of broken basaltic material, filling it with hydrogen chloride dissolved in water so that the acid becomes neutralized, and then carbonating the solution by bubbling carbon dioxide through it. Oxygen would not be permitted into the system. The hydrochloric acid would leach the rocks, resulting in the release and precipitation of silica and the production of a chloride ocean containing sodium, potassium, calcium, magnesium, aluminum, iron, and reduced sulfur species in the proportions present in the rocks. As complete neutralization was approached, the aluminum could begin to precipitate as hydroxides and then combine with precipitated silica to form cation-deficient aluminosilicates. As the neutralization process reached its end, the aluminosilicates would combine with more silica and with cations to form such minerals as chlorite, and ferrous iron would combine with silica and sulfur to produce greenalite and pyrite. In the final solution, chlorine would be balanced by sodium and calcium in roughly equal proportions, with subordinate amounts of potassium and magnesium; aluminum would be quantitatively removed, and silicon would be at saturation with amorphous silica. If this solution were then carbonated, calcium would be removed as calcium carbonate, and the chlorine balance would be maintained by abstraction of more sodium from the primary rock. The sediments formed in this system would contain chiefly silica, ferrous iron silicates, chloritic minerals, calcium carbonate, calcium-magnesium carbonates, and small amounts of pyrite.

If the hydrogen chloride added were in excess of the carbon dioxide, the resultant oceans would have a high content of calcium chloride ($CaCl_2$), but with a pH still near neutrality. If the carbon dioxide added were in excess of the chlorine, calcium would be precipitated as carbonate until it reached a level roughly that of present-day ocean waters—namely, a few hundred parts per million.

If this newly created ocean were left undisturbed for several hundred million years, its waters would evaporate and be transported onto the continents (in the form of precipitation); streams would transport their loads into it. The sediments produced in this ocean would be uplifted and incorporated into the continents. The influence of the continental debris would gradually be felt and the pH might change somewhat. Iron would be oxidized out of the ferrous silicates to yield iron oxides, but the composition of the water would not vary substantially.

The primary minerals of igneous rocks are all mildly basic compounds. When these minerals react in excess with acids such as hydrogen chloride and carbon dioxide, they produce neutral or mildly alkaline solutions as well as a set of altered aluminosilicate and carbonate reaction products. It is improbable that seawater has changed through time from a solution approximately in equilibrium with these reaction products—i.e., with clay minerals and carbonates.

The Modern Hydrosphere

It is likely that the hydrosphere achieved its modern chemical characteristics about 1.5 billion to 2 billion years ago. The chemical and mineralogical compositions and the relative proportions of sedimentary rocks of this age differ little from their counterparts of the Paleozoic Era (from 541 million to 252 million years ago). Calcium sulfate deposits dated to the late Precambrian (about 1.6 billion to 541 million years ago) attest to the fact that the acid sulfur gases had been neutralized to sulfate by this time. Chemically precipitated ferric oxides in late Precambrian sedimentary rocks indicate available free oxygen, whatever its percentage. The chemistry and mineralogy of middle and late Precambrian shales are similar to those of Paleozoic shales. The carbon isotopic signature of carbonate rocks has been remarkably constant for more than three billion years, indicating exceptional stability in size and fluxes related to organic carbon. The sulfur isotopic signature of sulfur phases in rocks strongly suggests that the sulfur cycle involving heterotrophic bacterial reduction of sulfate was in operation 2.7 billion years ago. It therefore appears that continuous cycling of sediments similar to those of today has occurred for 1.5 billion to 2 billion years and that these sediments have controlled hydrospheric, and particularly oceanic, composition.

Rain and dark cloud.

It was once thought that the saltiness of the modern oceans simply represents the storage of salts (that is, compounds that form when part of an acid is replaced by a metal or something like a metal) derived from rock weathering and transported to the oceans by fluvial processes. With increasing knowledge of the age of Earth, however, it was soon realized that, at the present rate of delivery of salts to the ocean or even at much reduced rates, the total salt content and the mass of individual salts in the oceans could be attained in geologically short time intervals compared with the planet's age. The total mass of salt in the oceans can be accounted for at today's rates of stream delivery in about 12 million years. The mass of dissolved silica in ocean water can be doubled in just 20,000 years by the addition of stream-derived silica. To double the sodium content would take 70 million years. It then became apparent that the oceans are not merely an accumulator of salts. Rather, as water evaporates from the oceans, together with some salt, the salts introduced must be removed in the form of minerals deposited in marine sediments. Accordingly, the concept of the oceans as a chemical system changed from that of a simple accumulator to that of a steady-state system in which rates of inflow of materials equal rates of outflow. The steady-state concept permits influx to vary with time, but the inflow would be matched by nearly simultaneous and equal variation of efflux.

In such a steady-state conceptual view of the oceans, it has been found necessary to treat components of ocean water in terms of all their influxes and effluxes and to be more cognizant of the time scale of application of the steady-state concept. Indeed, the recent increase in the carbon dioxide concentration of the atmosphere due to the burning of fossil fuels has been linked to a changed pH and dissolved inorganic carbon concentrations of surface ocean water on a time scale measured in hundreds of years.

If fossil-fuel burning were to cease, return to the original state of seawater composition could take thousands of years. Ocean water is not in steady state with respect to carbon on these time scales, but on a longer geologic time scale it certainly could be. Even on this longer time scale, however, oceanic composition has varied because of natural changes in the carbon dioxide level of the atmosphere and because of other factors.

It appears that the best description of modern seawater composition is that of a chemical system in a dynamic quasi-steady state. Changes in composition may occur over time, but the system always seems to return to a time-averaged steady-state composition. In other words, since 1.5 billion to 2 billion years ago, evolutionary chemical changes in the hydrosphere have been small when viewed against the magnitude of previous change.

It should be noted that rivers supply dissolved constituents to the oceans, whereas high and low temperature reactions between seawater and submarine basalts and reactions in sediment pore waters may add or remove constituents from ocean water. Biological processes involved in the formation of the opaline silica skeletons of diatoms (a type of algae with a shell made of silica) and radiolarians (planktonic protozoans) and the carbonate skeletons of planktonic foraminiferans (shelled unicellular organisms possessing pseudopods) and coccolithophorids (marine planktonic biflagellated organisms that secrete a type of minute calcium carbonate platelet or ring) chiefly remove calcium and silica from seawater. Exchange reactions between river-borne clays entering seawater are particularly significant for sodium and calcium ions. Most of the carbon imbalance in ocean water represents carbon released to the ocean-atmosphere system during precipitation of carbonate minerals; i.e.,

$$Ca^{2+} + 2HCO_3^- \rightarrow CaCO_3 + CO_2 + H_2O.$$

In the case of iron, it has been documented that "dissolved" iron carried by rivers is rapidly precipitated as hydroxides in the mixing zone with seawater and that the reduced dissolved iron released from anaerobic sediments also is rapidly precipitated under the oxic conditions (i.e., those with oxygen present) prevailing in the water column. Iron is also precipitated as iron smectites, hydrated iron oxides, and nontronite (iron-rich montmorillonite) in the deep sea. It is thus likely that iron is removed by these processes.

Sodium Imbalance

The imbalance in sodium is large: 45 percent of the river input is not accounted for in the mass balance calculations. There are, however, major uncertainties in the estimation of the pore-water flux of sodium ions. An important sink for sodium on a geologic time scale is the formation of evaporites. If the amount of unbalanced sodium is expressed in terms of halite deposition, it would correspond to 1.6×10^{14} grams of sodium chloride per year as compared with a potential total depositional rate of

3.3×10^{14} grams annually. There are no important sodium chloride deposits forming today. Thus, one possibility is that sodium is accumulating in the oceans. If so, in 6×10^6 years at an accumulation rate of 63×10^{12} grams of sodium annually, the average salinity of the oceans would increase less than one part per thousand. The chlorine balance for the oceans, however, indicates that it is likely that the major problem in the imbalance for sodium lies in the flux estimates for sediment pore waters and perhaps submarine weathering processes.

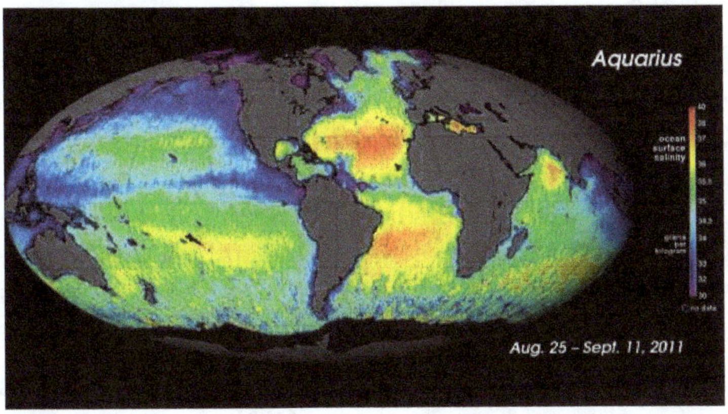

The first map of ocean salinity taken by the Aquarius/SAC-D spacecraft.

Modern seawater chemistry has been characteristic of roughly the past 600 million years of ocean history. Evaporite sediments provide strong evidence that the composition of seawater has not varied a great deal during this interval of geologic time. Nonetheless, it seems likely that fluctuations did occur, particularly in the concentrations of calcium, magnesium, and sulfate ions. The isotopic composition of sulfur in seawater, as recorded in evaporites, has varied dramatically during the past one billion years. Although it is difficult to relate these isotopic fluctuations to the calcium and sulfate concentrations of seawater, some scientists believe that the fluctuations do in fact imply changes in the latter. Furthermore, the major features of the sulfur isotopic curve for evaporites versus Phanerozoic time (the last 541 million years) is similar to that of the strontium-87/strontium-86 ratio, and perhaps the strontium/calcium ratio, of sedimentary materials during this time interval. Such covariation is consistent with a model in which fluxes related to alteration of seafloor basalts and continental river runoff vary with time, resulting in variation in seawater composition.

Changes in the Chemistry of the Atmosphere-Hydrosphere

The chemistry of the atmosphere has certainly changed significantly during the past one billion years of Earth history. A modification of this kind implies changes in the chemistry of the hydrosphere as well. Oxygen in the atmosphere rose substantially between two billion years ago and the beginning of the Phanerozoic Eon (i.e., 541 million years ago), whereas atmospheric carbon dioxide levels probably decreased. This change led in general to a progressively more oxygenated and less acidic hydrosphere.

It is likely that the development of higher land plants during the Devonian Period (from 419 million to about 359 million years ago) resulted in an increase in atmospheric oxygen and a decrease in carbon dioxide. Air trapped within bubbles in Arctic and Antarctic ice shows that the carbon dioxide content of the atmosphere during the climax of the last ice age was about 180 parts per million by volume (ppmv), and atmospheric CO_2 levels reached approximately 280 ppmv during the last great interglacial of 120,000 years ago, long before modern society initiated its extensive fossil-fuel burning and deforestation activities.

These atmospheric changes in themselves can influence the chemistry of the hydrosphere, but they also appear to be coupled with other changes in the rock-ocean-atmosphere-biota system that strongly affect hydrospheric chemistry. For example, though surface waters probably remained oxygenated during the Cretaceous and Devonian periods, there is evidence that intermediate and deep ocean waters were more anoxic (oxygen-depleted) than today. These "anoxic events" are characterized by dramatic changes in Earth's ocean-atmosphere system, including changes in the rates of the cyclic transfer of elements at Earth's surface and in atmospheric composition.

The probable impact of a bolide (either an asteroid or comet) and increased volcanism at the end of the Cretaceous (about 66 million years ago), though still the subject of hot debate, certainly could have caused short-term changes in the chemistry of Earth's atmosphere-hydrosphere. Scientists speculate that such an event could have had any of several results, including (1) changes in Earth's radiant-energy balance because of vast amounts of particulate and gaseous input into the atmosphere and subsequent cooling, (2) acid rain stemming from the input into the atmosphere of nitrogen gases generated by the bolide impact, (3) increased trace metal fluxes to the hydrosphere brought on by the destruction of the bolide and increased volcanism, and perhaps (4) increased "anoxia" of the hydrosphere as a result of the death of land and marine organisms.

Whatever the case, evidence of change is present in the rock record, albeit the composition of the modern hydrosphere has not varied greatly for the past one billion years. Moreover, as will be seen in the following section, humans are modifying the chemistry not only of local and regional water bodies but also of the entire global atmosphere-hydrosphere by increasing the rates of input of natural substances and by introducing new synthetic substances to the environment.

Impact of Human Activities on the Hydrosphere

The activities of modern society are having a severe impact on the hydrologic cycle. The dynamic steady state is being disturbed by the discharge of toxic chemicals, radioactive substances, and other industrial wastes and by the seepage of mineral fertilizers, herbicides, and pesticides into surface and subsurface aquatic systems. Inadvertent and deliberate discharge of petroleum, improper sewage disposal, and thermal pollution also are seriously affecting the quality of the hydrosphere.

Eutrophication

Historically, aquatic systems have been classified as oligotrophic or eutrophic. Oligotrophic waters are poorly fed by the nutrients nitrogen and phosphorus and have low concentrations of these constituents. There is thus low production of organic matter by photosynthesis in such waters. By contrast, eutrophic waters are well supplied with nutrients and generally have high concentrations of nitrogen and phosphorus and, correspondingly, large concentrations of plankton owing to high biological productivity. The waters of such aquatic systems are usually murky, and lakes and coastal marine systems may be oxygen-depleted at depth. The process of eutrophication is defined as high biological productivity resulting from increased input of nutrients or organic matter into aquatic systems. For lakes, this increased biological productivity usually leads to decreased lake volume because of the accumulation of organic detritus. Natural eutrophication occurs as aquatic systems fill in with organic matter; it is distinct from cultural eutrophication, which is caused by human intervention. The latter is characteristic of aquatic systems that have been artificially enriched by excess nutrients and organic matter from sewage, agriculture, and industry. Naturally eutrophic lakes may produce 75–250 grams of carbon per square metre per year, whereas those lakes experiencing eutrophication because of human activities can support 75–750 grams per square metre per year. Commonly, culturally eutrophic aquatic systems may exhibit extremely low oxygen concentrations in bottom waters. This is particularly true of stratified systems, as, for instance, lakes during summer where concentrations of molecular oxygen may reach levels of less than about one milligram per litre—a threshold for various biological and chemical processes.

Aquatic systems may change from oligotrophic to eutrophic, or the rate of eutrophication of a natural eutrophic system may be accelerated by the addition of nutrients and organic matter due to human activities. The process of cultural eutrophication, however, can be reversed if the excess nutrient and organic matter supply is shut off.

Not only do freshwater aquatic systems undergo cultural eutrophication, but coastal marine systems also may be affected by this process. On a global scale, the input by rivers of organic matter to the oceans today is twice the input in prehuman times, and the flux of nitrogen, together with that of phosphorus, has more than doubled. This excess loading of carbon, nitrogen, and phosphorus is leading to cultural eutrophication of marine systems. In several polluted eastern U.S. estuaries (e.g., Chesapeake and Delaware bays) and in some estuaries of western Europe (e.g., the Scheldt of Belgium and the Netherlands), all of the dissolved silica brought into the estuarine waters by rivers is removed by phytoplankton growth (primarily diatoms) resulting from excess fluxes of nutrients and organic matter. In the North Sea there is now a deficiency of silica and an excess of nitrogen and phosphorus, which in turn has led to a decrease in diatom productivity and an increase in cyanobacteria productivity—a biotic change brought about by cultural eutrophication.

Acid Rain

The emission of sulfur dioxide and nitrogen oxides to the atmosphere by human activities—primarily fossil-fuel burning—has led to the acidification of rain and freshwater aquatic systems. Acid rain is a worldwide problem and has been well documented for eastern North America and the countries of western Europe.

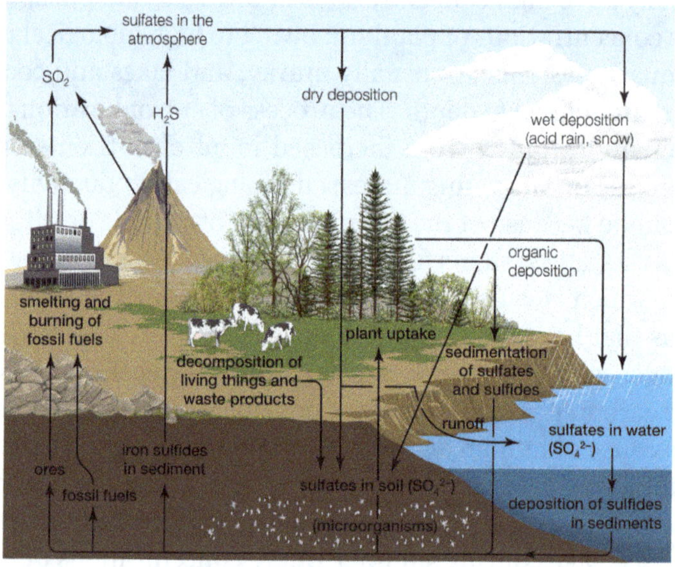

Sulfur cycle.

Acid rain is defined as precipitation with a pH of less than 5.2 that results from reactions involving gases other than carbon dioxide. The overall reactions that produce such precipitation are:

$$SO_2 + OH \rightarrow \text{intermediate species} \rightarrow H_2SO_4$$

$$SO_2 + H_2O_2 = H_2SO_4.$$

$$H_2SO_4 = 2H^+ + SO_4^{2-}.$$

$$NO_2 + OH \rightarrow HNO_3.$$

Average pH can be calculated as the −log aH^+ (aH^+ is activity of the hydrogen ion). A water chemistry study examining precipitation over the eastern United States for the period October 1979 through September 1980 revealed that low pH values were a result of equilibration of rainwater with the atmospheric acid gases of carbon, nitrogen, and sulfur. Equilibration only with atmospheric carbon dioxide would give a pH of 5.7. The significantly lower values are a result of reactions with nitrogen- and sulfur-bearing gaseous atmospheric components derived primarily from fossil-fuel burning sources.

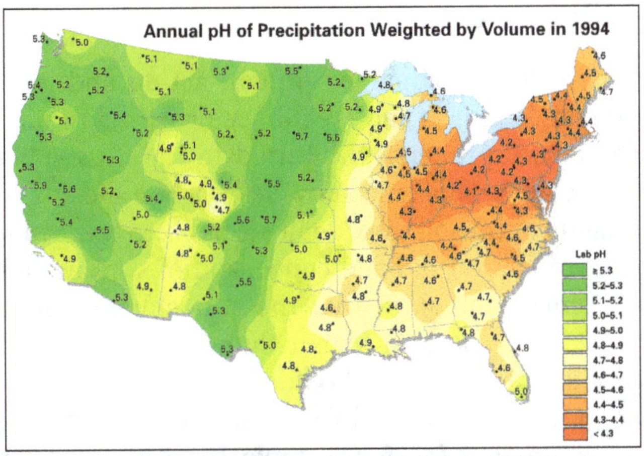

Map of precipitation pH in the continental United States in 1994.

Nitrate and sulfate concentrations in precipitation over the eastern United States are strongly correlated with pH—the lower the pH of rain, the higher the concentrations of nitrate and sulfate. Such low pH values and increased nitrate and sulfate concentrations were observed in the rains of western Europe and North America until the late 20th century. The pH values of precipitation in these regions have increased significantly since then because of strict air quality regulations. Other parts of the world that have industrialized since the late 20th century without enacting adequate air pollution controls, such as China, experienced similar pH declines in precipitation.

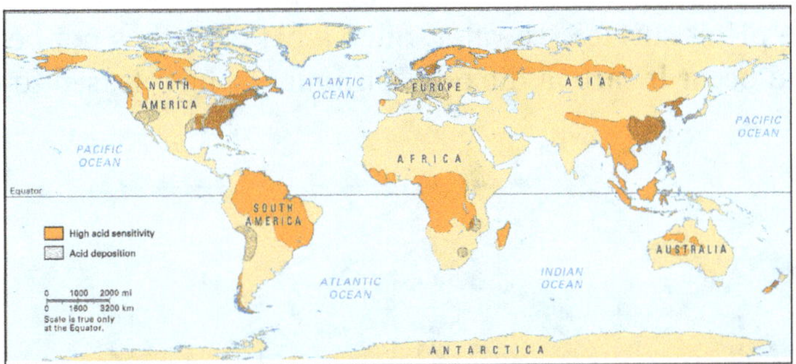

Areas affected by acid deposition contrasted with regions of high acid sensitivity.

Wet and dry deposition also removes the hydrogen ion produced in the rain by the oxidation and hydrolysis of these acid gases. This excess hydrogen ion can bring about the acidification of freshwater aquatic systems, particularly those with little buffer capacity (e.g., lakes situated in crystalline rock terrains). Furthermore, the lower pH values of rainwater, and consequently of soil water, can lead to increased mobilization of aluminum. Acidification of freshwater lakes in the eastern United States and southeastern Canada and increased aluminum concentrations in their waters are thought to be responsible for major changes in the ecosystems of the lakes. In particular, many lakes of this region lack substantial fish populations today, even though they supported large numbers of fish in the early 1900s. Acid rain also may be among the factors responsible for damage to the major forests and soils of the eastern United States and western Europe.

Buildup of Greenhouse Gases

One problem that was brought about by human action and is definitely affecting the hydrosphere globally is that of the greenhouse gases (so called because of their heat-trapping "greenhouse" properties) emitted to the atmosphere. Of the greenhouse gases released by anthropogenic activities, carbon dioxide has received much attention. Measurements of carbon dioxide in air bubbles trapped in ice and the continuous measurement of carbon dioxide concentrations in air samples collected at Mauna Loa, Hawaii, since 1958 show that the atmospheric concentration of more than 400 ppmv is roughly 45 percent higher than its late 1700s value of 275 ppmv. Much of this increase is due to carbon dioxide released to the atmosphere from the burning of coal, oil, gas, and wood and from the slash-and-burn activities that accompany deforestation practices (as, for example, those adopted in the Amazon River basin). The component of the hydrosphere most greatly affected by this emission of carbon dioxide is the ocean.

Before human activities had substantially affected the carbon dioxide cycle, there was a net flux of carbon dioxide from the oceans through the atmosphere to the land, where the gas was used in the net production of organic matter and the chemical weathering of minerals in continental rocks. Because of fossil-fuel burning and land-use practices, the net transfer from the ocean to the land has been reversed, and the ocean has now become an important sink of carbon dioxide. The net chemical reaction of adding carbon dioxide to the ocean (provided there is no reaction with carbonate solids) is and a lowering of the pH of surface seawater. Such a pH effect has not been observed but conceivably could occur if carbon dioxide continues to be released to the atmosphere by human activities.

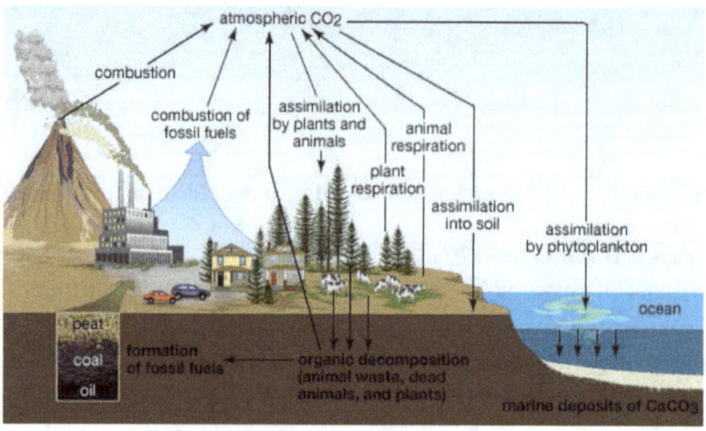

The generalized carbon cycle.

$$CO_2 + H_2O + CO_3^{2-} \text{ (carbonate ion)} = 2\,HCO_3^-$$

Based on greenhouse climate models and other considerations, it is possible that atmospheric carbon dioxide concentrations may double from their late 1700s level of

275 ppmv by the years 2030–50. By 2015, atmospheric carbon dioxide concentrations had surpassed 400 ppmv for the first time in 800,000 years. Climate models, which also consider the long-term warming potential made by other greenhouse gases (e.g., methane and nitrous oxide) in addition to that of carbon dioxide, project a rise in global mean surface temperature of 0.3 to 4.8 °C (0.5 to 8.6 °F) by 2100. This projected temperature increase would be two to three times greater at the poles than at the Equator and greater in the Arctic than in the Antarctic. At present there is no systematic worldwide program to decrease greenhouse gas emissions, except for that affecting chlorofluorocarbon (Freon) releases. Thus, it is conceivable that atmospheric carbon dioxide concentrations in the late 21st and early 22nd centuries might reach levels greater than twice their 1700s value. Whatever the case, the effect of the potential rise in surface temperature would be to speed up the hydrologic cycle and probably the rate of chemical weathering of continental rocks. Increases in the global mean evaporation and precipitation rates are expected from a doubling of the carbon dioxide level and a few degrees rise in global mean temperature. The effect on the water balance would be regional in nature, with some places becoming wetter and others drier. In general, there would be a trend toward greater and longer periods of summer dryness induced by lower soil moisture content and higher evaporation rates in the mid-latitudes of the Northern Hemisphere. In the arid western regions of the United States, which depend on irrigation for growing plants, severe water shortages could occur. By contrast, precipitation and runoff might increase, except in summer, at latitudes beyond 60° N because of a greater poleward transport of moisture. In summer, in a zone centred around 60° N, greater dryness might occur as a result of an earlier end of snowmelt.

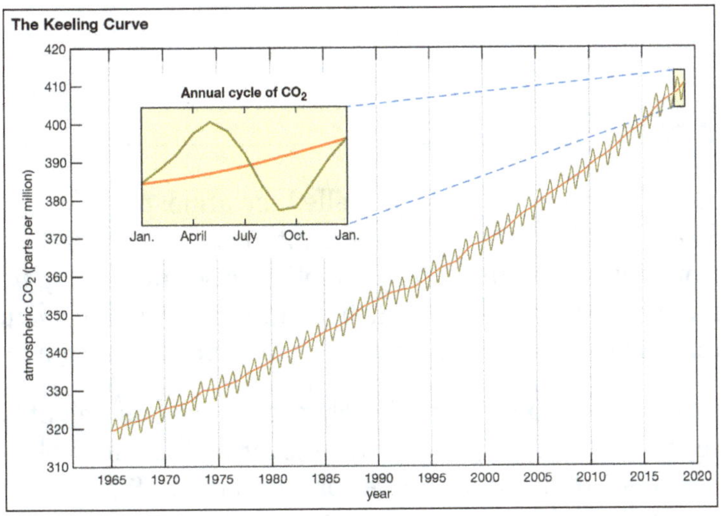

The Keeling Curve, named after American climate scientist Charles David Keeling, tracks changes in the concentration of carbon dioxide (CO_2) in Earth's atmosphere at a research station on Mauna Loa in Hawaii. Although these concentrations experience small seasonal fluctuations, the overall trend shows that CO_2 is increasing in the atmosphere.

Global warming could further affect the hydrologic cycle by the melting of ice and snow in the Greenland and Antarctic ice caps and in mountain glaciers, resulting in the transfer of water to the oceans. This process, together with thermal expansion of the oceans because of global warming, resulted in a slow rise in sea level during the 20th century, and it is expected to continue throughout the 21st century. If the West Antarctic Ice Sheet were to disintegrate, a much larger and more rapid rise in sea level of more than 3 metres (nearly 10 feet) could occur over the next several hundred years. The melting of all glacial ice would raise the sea level more than 66 metres (about 216 feet).

Significant reductions in the areal extent and thickness of sea ice in the Arctic occurring during the early 21st century have been attributed to global warming. Complete melting of the Arctic sea ice might occur, causing a northward shift in storm tracks and a reduction in Northern Hemispheric precipitation during the spring and fall. Furthermore, a worldwide reduction in sea ice might lead to increased evaporation from the ocean and increased low-altitude cloudiness, which would reflect solar radiation and cause cooling.

The potential changes in the hydrologic cycle induced by global warming resulting from anthropogenic emissions of greenhouse gases do not seem great. Yet, their consequences could be severe for ecosystems and human populations, especially since the latter are so sensitive to and dependent on such changes. A global rise in sea level of 1 metre (3.3 feet), for example, would almost completely inundate the coastal areas of Bangladesh. Agricultural lands could be displaced, just as patterns of arid, semiarid, and wet lands might become modified. It is essential that society plan for such potential changes so that, if they do occur, appropriate adjustments can be made to ameliorate them.

ATMOSPHERE

We live at the bottom of an invisible ocean called the atmosphere, a layer of gases surrounding our planet. Nitrogen and oxygen account for 99 percent of the gases in dry air, with argon, carbon dioxide, helium, neon, and other gases making up minute portions. Water vapor and dust are also part of Earth's atmosphere. Other planets and moons have very different atmospheres, and some have no atmospheres at all.

The atmosphere is so spread out that we barely notice it, yet its weight is equal to a layer of water more than 10 meters (34 feet) deep covering the entire planet. The bottom 30 kilometers (19 miles) of the atmosphere contains about 98 percent of its mass. The atmosphere—air—is much thinner at high altitudes. There is no atmosphere in space.

Scientists say many of the gases in our atmosphere were ejected into the air by early volcanoes. At that time, there would have been little or no free oxygen surrounding the Earth. Free oxygen consists of oxygen molecules not attached to another element, like carbon (to form carbon dioxide) or hydrogen (to form water).

Free oxygen may have been added to the atmosphere by primitive organisms, probably bacteria, during photosynthesis. Photosynthesis is the process a plant or other autotroph uses to make food and oxygen from carbon dioxide and water. Later, more complex forms of plant life added more oxygen to the atmosphere. The oxygen in today's atmosphere probably took millions of years to accumulate.

The atmosphere acts as a gigantic filter, keeping out most ultraviolet radiation while letting in the sun's warming rays. Ultraviolet radiation is harmful to living things, and is what causes sunburns. Solar heat, on the other hand, is necessary for all life on Earth.

Earth's atmosphere has a layered structure. From the ground toward the sky, the layers are the troposphere, stratosphere, mesosphere, thermosphere, and exosphere. Another layer, called the ionosphere, extends from the mesosphere to the exosphere. Beyond the exosphere is outer space. The boundaries between atmospheric layers are not clearly defined, and change depending on latitude and season.

Troposphere

The troposphere is the lowest atmospheric layer. On average, the troposphere extends from the ground to about 10 kilometers (6 miles) high, ranging from about 6 kilometers (4 miles) at the poles to more than 16 kilometers (10 miles) at the Equator. The top of the troposphere is higher in summer than in winter.

Almost all weather develops in the troposphere because it contains almost all of the atmosphere's water vapor. Clouds, from low-lying fog to thunderheads to high-altitude cirrus, form in the troposphere. Air masses, areas of high-pressure and low-pressure systems, are moved by winds in the troposphere. These weather systems lead to daily weather changes as well as seasonal weather patterns and climate systems, such as El Nino.

Air in the troposphere thins as altitude increases. There are fewer molecules of oxygen at the top of Mount Everest, Nepal, for example, than there are on a beach in Hawaii. This is why mountaineers often use canisters of oxygen when climbing tall peaks. Thin air is also why helicopters have difficulty maneuvering at high altitudes. In fact, a helicopter was not able to land on Mount Everest until 2005.

As air in the troposphere thins, temperature decreases. This is why mountaintops are usually much colder than the valleys beneath. Scientists used to think temperature continued to drop as altitude increased beyond the troposphere. But data collected with weather balloons and rockets have showed this is not the case. In the lower stratosphere, temperature stays almost constant. As altitude increases in the stratosphere, temperature actually increases.

Solar heat penetrates the troposphere easily. This layer also absorbs heat that is reflected back from the ground in a process called the greenhouse effect. The greenhouse effect is necessary for life on Earth. The atmosphere's most abundant greenhouse gases are carbon dioxide, water vapor, and methane.

Fast-moving, high-altitude winds called jet streams swirl around the planet near the upper boundary of the troposphere. Jet streams are extremely important to the airline industry. Aircraft save time and money by flying in jet streams instead of the lower troposphere, where air is thicker.

Stratosphere

The troposphere tends to change suddenly and violently, but the stratosphere is calm. The stratosphere extends from the tropopause, the upper boundary of the troposphere, to about 50 kilometers (32 miles) above the Earth's surface.

Strong horizontal winds blow in the stratosphere, but there is little turbulence. This is ideal for planes that can fly in this part of the atmosphere.

The stratosphere is very dry and clouds are rare. Those that do form are thin and wispy. They are called nacreous clouds. Sometimes they are called mother-of-pearl clouds because their colors look like those inside a mollusk shell.

The stratosphere is crucial to life on Earth because it contains small amounts of ozone, a form of oxygen that prevents harmful UV rays from reaching Earth. The region within the stratosphere where this thin shell of ozone is found is called the ozone layer. The stratosphere's ozone layer is uneven, and thinner near the poles. The amount of ozone in the Earth's atmosphere is declining steadily. Scientists have linked use of chemicals such as chlorofluorocarbons (CFCs) to ozone depletion.

Mesosphere

The mesosphere extends from the stratopause (the upper boundary of the stratosphere) to about 85 kilometers (53 miles) above the surface of the Earth. Here, temperatures again begin to fall.

The mesosphere has the coldest temperatures in the atmosphere, dipping as low as -120 degrees Celsius (-184 degrees Fahrenheit, or 153 kelvin). The mesosphere also has the atmosphere's highest clouds. In clear weather, you can sometimes see them as silvery wisps immediately after sunset. They are called noctilucent clouds, or night-shining clouds. The mesosphere is so cold that noctilucent clouds are actually frozen water vapor ice clouds.

Shooting stars the fiery burnout of meteors, dust, and rocks from outer space are visible in the mesosphere. Most shooting stars are the size of a grain of sand and burn up before entering the stratosphere or troposphere. However, some meteors are the size of pebbles or even boulders. Their outer layers burn as they race through the mesosphere, but they are massive enough to fall through the lower atmosphere and crash to Earth as meteorites.

The mesosphere is the least understood part of Earth's atmosphere. It is too high for aircraft or weather balloons to operate, but too low for spacecraft. Sounding rockets have provided meteorologists and astronomers their only significant data on this important part of the atmosphere. Sounding rockets are unmanned research instruments that collect data during sub-orbital flights.

Perhaps because the mesosphere is so little understood, it is home to two meteorological mysteries: sprites and elves. Sprites are reddish, vertical electrical discharges that appear high above thunderheads, in the upper stratosphere and mesosphere. Elves are dim, halo-shaped discharges that appear even higher in the mesosphere.

Ionosphere

The ionosphere extends from the top half of the mesosphere all the way to the exosphere. This atmospheric layer conducts electricity.

The ionosphere is named for ions created by energetic particles from sunlight and outer space. Ions are atoms in which the number of electrons does not equal the number of protons, giving the atom a positive (fewer electrons than protons) or negative (more electrons than protons) charge. Ions are created as powerful X-rays and UV rays knock electrons off atoms.

The ionosphere a layer of free electrons and ions reflects radio waves. Guglielmo Marconi, the "Father of Wireless," helped prove this in 1901 when he sent a radio signal from Cornwall, England, to St. John's, Newfoundland, Canada. Marconi's experiment demonstrated that radio signals did not travel in a straight line, but bounced off an atmospheric layer the ionosphere.

The ionosphere is broken into distinct layers, called the D, E, F1, and F2 layers. Like all other parts of the atmosphere, these layers vary with season and latitude. Changes in the ionosphere actually happen on a daily basis. The low D layer, which absorbs high-frequency radio waves, and the E layer actually disappear at night, which means radio waves can reach higher into the ionosphere. That's why AM radio stations can extend their range by hundreds of kilometers every night.

The ionosphere also reflects particles from solar wind, the stream of highly charged particles ejected by the sun. These electrical displays create auroras (light displays) called the Northern and Southern Lights.

Thermosphere

The thermosphere is the thickest layer in the atmosphere. Only the lightest gases mostly oxygen, helium, and hydrogen are found here.

The thermosphere extends from the mesopause (the upper boundary of the mesosphere) to 690 kilometers (429 miles) above the surface of the Earth. Here, thinly scattered molecules of gas absorb x-rays and ultraviolet radiation. This absorption process propels the molecules in the thermosphere to great speeds and high temperatures. Temperatures in the thermosphere can rise to 1,500 degrees Celsius (2,732 degrees Fahrenheit, or 1,773 kelvin).

Though the temperature is very high, there is not much heat. How is that possible? Heat is created when molecules get excited and transfer energy from one molecule to another. Heat happens in an area of high pressure (think of water boiling in a pot). Since there is very little pressure in the thermosphere, there is little heat transfer.

The Hubble Space Telescope and the International Space Station (ISS) orbit the Earth in the thermosphere. Even though the thermosphere is the second-highest layer of Earth's atmosphere, satellites that operate here are in "low-Earth orbit."

Exosphere

The fluctuating area between the thermosphere and the exosphere is called the turbopause. The lowest level of the exosphere is called the exobase. At the upper boundary of the exosphere, the ionosphere merges with interplanetary space, or the space between planets.

The exosphere expands and contracts as it comes into contact with solar storms. In solar storms particles are flung through space from explosive events on the sun, such as solar flares and Coronal Mass Ejections (CMEs).

Solar storms can squeeze the exosphere to just 1,000 kilometers (620 miles) above the Earth. When the sun is calm, the exosphere can extend 10,000 kilometers (6,214 miles).

Hydrogen, the lightest element in the universe, dominates the thin atmosphere of the exosphere. Only trace amounts of helium, carbon dioxide, oxygen, and other gases are present.

Many weather satellites orbit Earth in the exosphere. The lower part of the exosphere includes low-Earth orbit, while medium-Earth orbit is higher in the atmosphere.

The upper boundary of the exosphere is visible in satellite images of Earth, called the geocorona, it is the fuzzy blue illumination that circles the Earth.

References

- Biosphere, science: britannica.com, Retrieved 12 July, 2019
- Lithosphere, encyclopedia: nationalgeographic.org, Retrieved 11 January, 2019
- Geosphere, geo, mtpe: earth.rice.edu, Retrieved 1 August, 2019
- Hydrosphere, science: britannica.com, Retrieved 5 March, 2019
- Atmosphere, encyclopedia: nationalgeographic.org, Retrieved 8 June, 2019

2

Biosphere of the Earth

Biosphere can be considered to be the worldwide sum of all ecosystems, which incorporates all the living beings and their relationships with each other. The diverse aspects related to the Earth's biosphere such as the significance of water in it have been thoroughly discussed in this chapter.

The biosphere is made up of the parts of Earth where life exists. The biosphere extends from the deepest root systems of trees, to the dark environment of ocean trenches, to lush rain forests and high mountaintops.

The solid surface layer of the Earth is the lithosphere. The atmosphere is the layer of air that stretches above the lithosphere. The Earth's water—on the surface, in the ground, and in the air—makes up the hydrosphere.

Since life exists on the ground, in the air, and in the water, the biosphere overlaps all these spheres. Although the biosphere measures about 20 kilometers (12 miles) from top to bottom, almost all life exists between about 500 meters (1,640 feet) below the ocean's surface to about 6 kilometers (3.75 miles) above sea level.

ECOSYSTEM

An ecosystem is a community of living organisms in conjunction with the nonliving components of their environment, interacting as a system. These biotic and abiotic components are linked together through nutrient cycles and energy flows. Energy enters the system through photosynthesis and is incorporated into plant tissue. By feeding on plants and on one-another, animals play an important role in the movement of matter and energy through the system. They also influence the quantity of plant and microbial biomass present. By breaking down dead organic matter, decomposers release carbon back to the atmosphere and facilitate nutrient cycling by converting nutrients stored in dead biomass back to a form that can be readily used by plants and other microbes.

Ecosystems are controlled by external and internal factors. External factors such as climate, parent material which forms the soil and topography, control the overall structure of an ecosystem but are not themselves influenced by the ecosystem. Unlike external factors, internal factors are controlled, for example, decomposition, root competition, shading, disturbance, succession, and the types of species present.

Ecosystems are dynamic entities—they are subject to periodic disturbances and are in the process of recovering from some past disturbance. Ecosystems in similar environments that are located in different parts of the world can end up doing things very differently simply because they have different pools of species present. Internal factors not only control ecosystem processes but are also controlled by them and are often subject to feedback loops.

Resource inputs are generally controlled by external processes like climate and parent material. Resource availability within the ecosystem is controlled by internal factors like decomposition, root competition or shading. Although humans operate within ecosystems, their cumulative effects are large enough to influence external factors like climate.

Biodiversity affects ecosystem functioning, as do the processes of disturbance and succession. Ecosystems provide a variety of goods and services upon which people depend.

Processes

Rainforest ecosystems are rich in biodiversity. This is the Gambia River in Senegal's Niokolo-Koba National Park.

Flora of Baja California Desert, Cataviña region, Mexico.

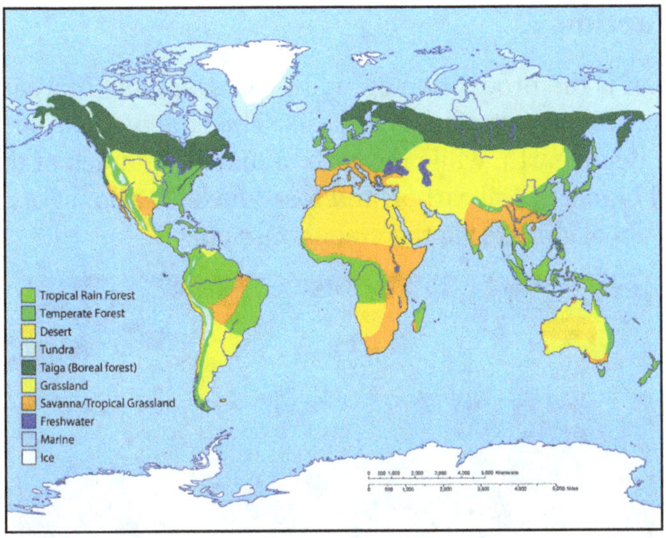

Biomes of the world.

Ecosystems are controlled both by external and internal factors. External factors, also called state factors, control the overall structure of an ecosystem and the way things work within it, but are not themselves influenced by the ecosystem. The most important of these is climate. Climate determines the biome in which the ecosystem is embedded. Rainfall patterns and seasonal temperatures influence photosynthesis and thereby determine the amount of water and energy available to the ecosystem.

Parent material determines the nature of the soil in an ecosystem, and influences the supply of mineral nutrients. Topography also controls ecosystem processes by affecting things like microclimate, soil development and the movement of water through a system. For example, ecosystems can be quite different if situated in a small depression on the landscape, versus one present on an adjacent steep hillside.

Other external factors that play an important role in ecosystem functioning include time and potential biota. Similarly, the set of organisms that can potentially be present in an area can also significantly affect ecosystems. Ecosystems in similar environments that are located in different parts of the world can end up doing things very differently simply because they have different pools of species present. The introduction of non-native species can cause substantial shifts in ecosystem function.

Unlike external factors, internal factors in ecosystems not only control ecosystem processes but are also controlled by them. Consequently, they are often subject to feedback loops. While the resource inputs are generally controlled by external processes like climate and parent material, the availability of these resources within the ecosystem is controlled by internal factors like decomposition, root competition or shading. Other factors like disturbance, succession or the types of species present are also internal factors.

Primary Production

Primary production is the production of organic matter from inorganic carbon sources. This mainly occurs through photosynthesis. The energy incorporated through this process supports life on earth, while the carbon makes up much of the organic matter in living and dead biomass, soil carbon and fossil fuels. It also drives the carbon cycle, which influences global climate via the greenhouse effect.

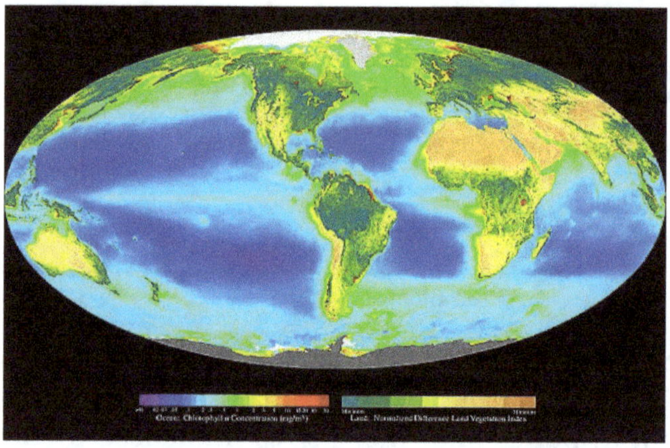

Global oceanic and terrestrial phototroph abundance, from September 1997 to August 2000. As an estimate of autotroph biomass, it is only a rough indicator of primary production potential and not an actual estimate of it.

Through the process of photosynthesis, plants capture energy from light and use it to combine carbon dioxide and water to produce carbohydrates and oxygen. The photosynthesis carried out by all the plants in an ecosystem is called the gross primary production (GPP). About half of the GPP is consumed in plant respiration. The remainder, that portion of GPP that is not used up by respiration, is known as the net primary production (NPP). Total photosynthesis is limited by a range of environmental factors. These include the amount of light available, the amount of leaf area a plant has to capture light (shading by other plants is a major limitation of photosynthesis), rate at which carbon dioxide can be supplied to the chloroplasts to support photosynthesis, the availability of water, and the availability of suitable temperatures for carrying out photosynthesis.

Energy Flow

Energy and carbon enter ecosystems through photosynthesis, are incorporated into living tissue, transferred to other organisms that feed on the living and dead plant matter, and eventually released through respiration.

The carbon and energy incorporated into plant tissues (net primary production) is either consumed by animals while the plant is alive, or it remains uneaten when the plant tissue dies and becomes detritus. In terrestrial ecosystems, roughly 90% of the net primary production ends up being broken down by decomposers. The remainder is either consumed by animals while still alive and enters the plant-based trophic system, or it is consumed after it has died, and enters the detritus-based trophic system.

In aquatic systems, the proportion of plant biomass that gets consumed by herbivores is much higher. In trophic systems photosynthetic organisms are the primary producers. The organisms that consume their tissues are called primary consumers or secondary producers herbivores. Organisms which feed on microbes (bacteria and fungi) are termed microbivores. Animals that feed on primary consumers carnivores are secondary consumers. Each of these constitutes a trophic level.

The sequence of consumption from plant to herbivore, to carnivore forms a food chain. Real systems are much more complex than this organisms will generally feed on more than one form of food, and may feed at more than one trophic level. Carnivores may capture some prey which are part of a plant-based trophic system and others that are part of a detritus-based trophic system (a bird that feeds both on herbivorous grasshoppers and earthworms, which consume detritus). Real systems, with all these complexities, form food webs rather than food chains. The food chain usually consist of four level of consumption which are producers, primary consumers, secondary consumers, and tertiary consumers.

Decomposition

The carbon and nutrients in dead organic matter are broken down by a group of processes known as decomposition. This releases nutrients that can then be re-used for plant and microbial production and returns carbon dioxide to the atmosphere (or water) where it can be used for photosynthesis. In the absence of decomposition, the dead organic matter would accumulate in an ecosystem, and nutrients and atmospheric carbon dioxide would be depleted. Approximately 90% of terrestrial net primary production goes directly from plant to decomposer.

Decomposition processes can be separated into three categories leaching, fragmentation and chemical alteration of dead material. As water moves through dead organic matter, it dissolves and carries with it the water-soluble components. These are then taken up by organisms in the soil, react with mineral soil, or are transported beyond the confines of the ecosystem (and are considered lost to it). Newly shed leaves and newly dead animals have high concentrations of water-soluble components and include sugars, amino acids and mineral nutrients. Leaching is more important in wet environments and much less important in dry ones.

Fragmentation processes break organic material into smaller pieces, exposing new surfaces for colonization by microbes. Freshly shed leaf litter may be inaccessible due to an outer layer of cuticle or bark, and cell contents are protected by a cell wall. Newly dead animals may be covered by an exoskeleton. Fragmentation processes, which break through these protective layers, accelerate the rate of microbial decomposition. Animals fragment detritus as they hunt for food, as does passage through the gut. Freeze-thaw cycles and cycles of wetting and drying also fragment dead material.

The chemical alteration of the dead organic matter is primarily achieved through bacterial and fungal action. Fungal hyphae produce enzymes which can break through the tough outer structures surrounding dead plant material. They also produce enzymes which break down lignin, which allows them access to both cell contents and to the nitrogen in the lignin. Fungi can transfer carbon and nitrogen through their hyphal networks and thus, unlike bacteria, are not dependent solely on locally available resources.

Decomposition rates vary among ecosystems. The rate of decomposition is governed by three sets of factors—the physical environment (temperature, moisture, and soil properties), the quantity and quality of the dead material available to decomposers, and the nature of the microbial community itself. Temperature controls the rate of microbial respiration; the higher the temperature, the faster microbial decomposition occurs. It also affects soil moisture, which slows microbial growth and reduces leaching. Freeze-thaw cycles also affect decomposition—freezing temperatures kill soil microorganisms, which allows leaching to play a more important role in moving nutrients around. This can be especially important as the soil thaws in the spring, creating a pulse of nutrients which become available.

Decomposition rates are low under very wet or very dry conditions. Decomposition rates are highest in wet, moist conditions with adequate levels of oxygen. Wet soils tend to become deficient in oxygen (this is especially true in wetlands), which slows microbial growth. In dry soils, decomposition slows as well, but bacteria continue to grow (albeit at a slower rate) even after soils become too dry to support plant growth.

Nutrient Cycling

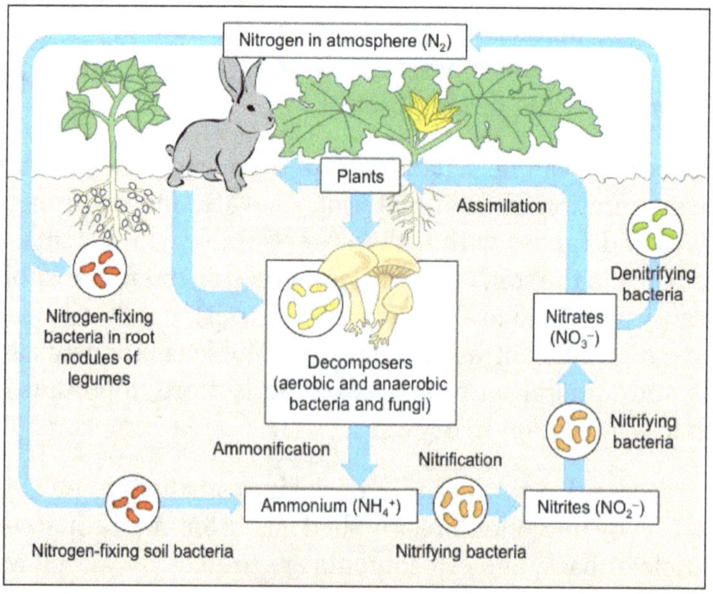

Biological nitrogen cycling.

Ecosystems continually exchange energy and carbon with the wider environment. Mineral nutrients, on the other hand, are mostly cycled back and forth between plants, animals, microbes and the soil. Most nitrogen enters ecosystems through biological nitrogen fixation, is deposited through precipitation, dust, gases or is applied as fertilizer.

Since most terrestrial ecosystems are nitrogen-limited, nitrogen cycling is an important control on ecosystem production.

Until modern times, nitrogen fixation was the major source of nitrogen for ecosystems. Nitrogen-fixing bacteria either live symbiotically with plants or live freely in the soil. The energetic cost is high for plants which support nitrogen-fixing symbionts as much as 25% of gross primary production when measured in controlled conditions. Many members of the legume plant family support nitrogen-fixing symbionts. Some cyanobacteria are also capable of nitrogen fixation. These are phototrophs, which carry out photosynthesis. Like other nitrogen-fixing bacteria, they can either be free-living or have symbiotic relationships with plants. Other sources of nitrogen include acid deposition produced through the combustion of fossil fuels, ammonia gas which evaporates from agricultural fields which have had fertilizers applied to them, and dust. Anthropogenic nitrogen inputs account for about 80% of all nitrogen fluxes in ecosystems.

When plant tissues are shed or are eaten, the nitrogen in those tissues becomes available to animals and microbes. Microbial decomposition releases nitrogen compounds from dead organic matter in the soil, where plants, fungi, and bacteria compete for it. Some soil bacteria use organic nitrogen-containing compounds as a source of carbon, and release ammonium ions into the soil. This process is known as nitrogen mineralization. Others convert ammonium to nitrite and nitrate ions, a process known as nitrification. Nitric oxide and nitrous oxide are also produced during nitrification. Under nitrogen-rich and oxygen-poor conditions, nitrates and nitrites are converted to nitrogen gas, a process known as denitrification.

Other important nutrients include phosphorus, sulfur, calcium, potassium, magnesium and manganese. Phosphorus enters ecosystems through weathering. As ecosystems age this supply diminishes, making phosphorus-limitation more common in older landscapes (especially in the tropics). Calcium and sulfur are also produced by weathering, but acid deposition is an important source of sulfur in many ecosystems. Although magnesium and manganese are produced by weathering, exchanges between soil organic matter and living cells account for a significant portion of ecosystem fluxes. Potassium is primarily cycled between living cells and soil organic matter.

Function and Biodiversity

Biodiversity plays an important role in ecosystem functioning. The reason for this is that ecosystem processes are driven by the number of species in an ecosystem, the exact nature of each individual species, and the relative abundance organisms within these species. Ecosystem processes are broad generalizations that actually take place through the actions of individual organisms. The nature of the organisms, the species, functional groups and trophic levels to which they belong, dictates the sorts of actions these individuals are capable of carrying out and the relative efficiency with which they do so.

Loch Lomond in Scotland forms a relatively isolated ecosystem. The fish community of this lake has remained stable over a long period until a number of introductions in the 1970s restructured it's food web.

Spiny forest at Ifaty, Madagascar, featuring various *Adansonia* (baobab) species, *Alluaudia procera* (Madagascar ocotillo) and other vegetation.

Ecological theory suggests that in order to coexist, species must have some level of limiting similarity—they must be different from one another in some fundamental way, otherwise one species would competitively exclude the other. Despite this, the cumulative effect of additional species in an ecosystem is not linear—additional species may enhance nitrogen retention, for example, but beyond some level of species richness, additional species may have little additive effect.

The addition (or loss) of species which are ecologically similar to those already present in an ecosystem tends to only have a small effect on ecosystem function. Ecologically distinct species, on the other hand, have a much larger effect. Similarly, dominant species have a large effect on ecosystem function, while rare species tend to have a small effect. Keystone species tend to have an effect on ecosystem function that is disproportionate to their abundance in an ecosystem. Similarly, an ecosystem engineer is any organism that creates, significantly modifies, maintains or destroys a habitat.

Dynamics

Ecosystems are dynamic entities. They are subject to periodic disturbances and are in the process of recovering from some past disturbance. When a perturbation occurs, an ecosystem responds by moving away from its initial state. The tendency of an ecosystem to remain close to its equilibrium state, despite that disturbance, is termed its resistance. On the other hand, the speed with which it returns to its initial state after disturbance is called its resilience. Time plays a role in the development of soil from bare rock and the recovery of a community from disturbance.

From one year to another, ecosystems experience variation in their biotic and abiotic environments. A drought, a colder than usual winter, and a pest outbreak all are short-term variability in environmental conditions. Animal populations vary from year to year, building up during resource-rich periods and crashing as they overshoot their food supply. These changes play out in changes in net primary production decomposition rates, and other ecosystem processes. Longer-term changes also shape ecosystem processes—the forests of eastern North America still show legacies of cultivation which ceased 200 years ago, while methane production in eastern Siberian lakes is controlled by organic matter which accumulated during the Pleistocene.

Disturbance also plays an important role in ecological processes. F. Stuart Chapin and co-authors define disturbance as "a relatively discrete event in time and space that alters the structure of populations, communities, and ecosystems and causes changes in resources availability or the physical environment". This can range from tree falls and insect outbreaks to hurricanes and wildfires to volcanic eruptions. Such disturbances can cause large changes in plant, animal and microbe populations, as well soil organic matter content. Disturbance is followed by succession, a "directional change in ecosystem structure and functioning resulting from biotically driven changes in resources supply."

The frequency and severity of disturbance determine the way it affects ecosystem function. A major disturbance like a volcanic eruption or glacial advance and retreat leave behind soils that lack plants, animals or organic matter. Ecosystems that experience such disturbances undergo primary succession. A less severe disturbance like forest fires, hurricanes or cultivation result in secondary succession and a faster recovery. More severe disturbance and more frequent disturbance result in longer recovery times.

A freshwater lake in Gran Canaria, an island of the Canary Islands. Clear boundaries make lakes convenient to study using an ecosystem approach.

Human Activities

Human activities are important in almost all ecosystems. Although humans exist and operate within ecosystems, their cumulative effects are large enough to influence external factors like climate.

Ecosystem Goods and Services

The High Peaks Wilderness Area in the 6,000,000-acre (2,400,000 ha) Adirondack Park is an example of a diverse ecosystem.

Ecosystems provide a variety of goods and services upon which people depend. Ecosystem goods include the "tangible, material products" of ecosystem processes such as food, construction material, medicinal plants. They also include less tangible items like tourism and recreation, and genes from wild plants and animals that can be used to improve domestic species.

Ecosystem services, on the other hand, are generally "improvements in the condition or location of things of value". These include things like the maintenance of hydrological cycles, cleaning air and water, the maintenance of oxygen in the atmosphere, crop pollination and even things like beauty, inspiration and opportunities for research. While material from the ecosystem had traditionally been recognized as being the basis for things of economic value, ecosystem services tend to be taken for granted.

Ecosystem Management

When natural resource management is applied to whole ecosystems, rather than single species, it is termed ecosystem management. Although definitions of ecosystem management abound, there is a common set of principles which underlie these definitions. A fundamental principle is the long-term sustainability of the production of goods and services by the ecosystem; "intergenerational sustainability is a precondition for management, not an afterthought".

While ecosystem management can be used as part of a plan for wilderness conservation, it can also be used in intensively managed ecosystems.

Threats caused by Humans

As human population and per capita consumption grow, so do the resource demands imposed on ecosystems and the effects of the human ecological footprint. Natural resources are vulnerable and limited. The environmental impacts of anthropogenic actions are becoming more apparent. Problems for all ecosystems include: environmental pollution, climate change and biodiversity loss. For terrestrial ecosystems further threats include air pollution, soil degradation, and deforestation. For aquatic ecosystems threats include also unsustainable exploitation of marine resources (for example overfishing of certain species), marine pollution, microplastics pollution, water pollution, the warming of oceans, and building on coastal areas.

Society is increasingly becoming aware that ecosystem services are not only limited but also that they are threatened by human activities. The need to better consider long-term ecosystem health and its role in enabling human habitation and economic activity is urgent. To help inform decision-makers, many ecosystem services are being assigned economic values, often based on the cost of replacement with anthropogenic alternatives. The ongoing challenge of prescribing economic value to nature, for example through biodiversity banking, is prompting transdisciplinary shifts in how we recognize and manage the environment, social responsibility, business opportunities, and our future as a species.

LIFE ON EARTH

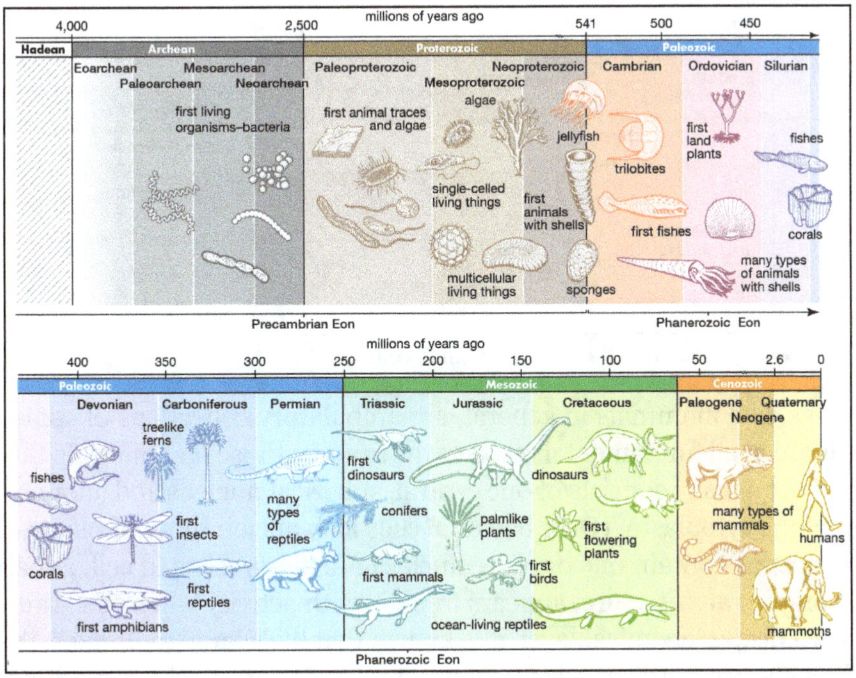

Over hundreds of millions of years, life spread through the seas and over Earth's surface. The first life-forms were small and simple. Later forms were more complicated and diverse.

The existence of diverse definitions of life surely means that life is complex and difficult to briefly define. A scientific understanding of living systems has existed since the second half of the 19th century. But the diversity of definitions and lack of consensus among professionals suggest something else as well. All organisms on Earth are extremely closely related, despite superficial differences. The fundamental pattern, both in form and in matter, of all life on Earth is essentially identical. This identity implies that all organisms on Earth are evolved from a single instance of the origin of life. To generalize from a single example is difficult, especially when the example itself is changing, growing, and evolving. In this respect the biologist is fundamentally handicapped, as compared with, say, the chemist, physicist, geologist, or meteorologist, each of whom can now study aspects of his discipline beyond Earth. If truly only one sort of life on Earth exists, then perspective is lacking in a most fundamental way. On the other hand, the historical continuity of all life-forms means that ancient life, perhaps even the origins of life, may be glimpsed by studying modern cells.

Chemistry of Life

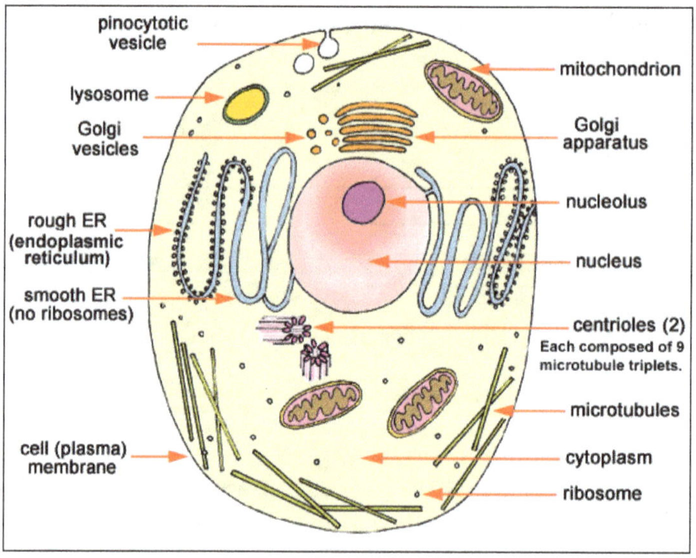

Animal cell.

Human beings, like mammals in general, are ambulatory collections of some 10^{14} cells. Human cells are in all fundamental respects the same as those that make up the other animals. Each cell typically consists of one central, spherical nucleus and another heterogeneous region, the cytoplasm. (Only bacterial cells lack nuclei; those of plants, fungi, and all other organisms contain one or more nuclei.) A living nucleated cell, a marvel of detailed and complex architecture, appears frenetic with activity when seen through a microscope. On a deeper chemical level, it is known that life's large molecules, the proteins and nucleic acids, are synthesized at a very fast rate. Enzymes, which speed up chemical reactions, are all proteins, but by no means are all proteins enzymes. An enzyme catalyzes the synthesis of more than 1,000 other molecules per second. The total mass of a metabolizing bacterial cell can be synthesized in 20 minutes. Although some feel debased by the implication that people are "nothing more" than a frenetic collection of interacting molecules, others are thrilled with the power of science to reveal the inner workings of the chemistry of life. The spectacular success of biochemistry and molecular biology in the 20th century suggests that laws of biology are derived from the interaction of atoms, thermodynamic principles, and life's chemistry, which has persisted with faithful continuity since its origin some 3.7 billion to 3.5 billion years ago.

DNA, RNA and Protein

The specific carrier of the genetic information in all organisms is the nucleic acid known as DNA, short for deoxyribonucleic acid. DNA is a double helix, two molecular coils wrapped around each other and chemically bound one to another by bonds connecting adjacent bases. Each long ladderlike DNA helix has a backbone that consists of a sequence of alternating sugars and phosphates. Attached to each sugar is a "base" consisting of the nitrogen-containing compound adenine, guanine, ctyosine, or thymine. Each sugar-phosphate-base "rung" is called a nucleotide. A very significant one-to-one pairing between bases occurs that ensures the connection of adjacent helices. Once the sequence of bases along one helix (half the ladder) has been specified, the sequence along the other half is also specified. The specificity of base pairing plays a key role in the replication of the DNA molecule. Each helix makes an identical copy of the other from molecular building blocks in the cell. These nucleic acid replication events are mediated by enzymes called DNA polymerases. With the aid of enzymes, DNA can be produced in the laboratory.

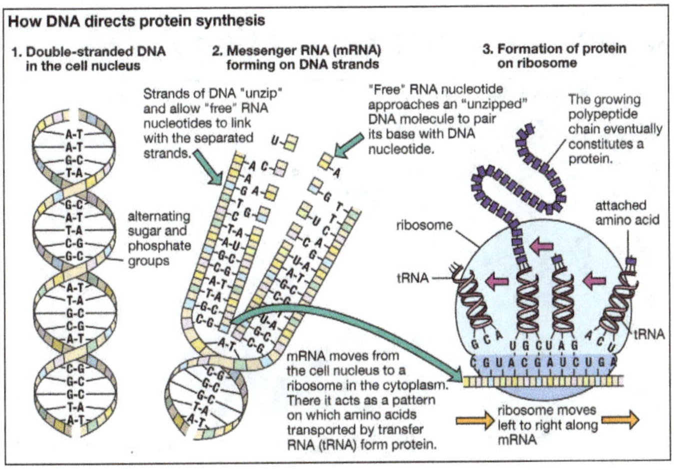

DNA in the cell nucleus carries a genetic code, which consists of sequences of adenine (A), thymine (T), guanine (G), and cytosine (C). RNA, which contains uracil (U) instead of thymine, carries the code to protein-making sites in the cell. To make RNA, DNA pairs its bases with those of the "free" nucleotides. Messenger RNA (mRNA) then travels to the ribosomes in the cell cytoplasm, where protein synthesis occurs. The base triplets of transfer RNA (tRNA) pair with those of mRNA and at the same time deposit their amino acids on the growing protein chain. Finally, the synthesized protein is released to perform its task in the cell or elsewhere in the body.

The cell, whether bacterial or nucleated, is the minimal unit of life. Many of the fundamental properties of cells are a function of their nucleic acids, their proteins, and the interactions among these molecules bounded by active membranes. Within the nuclear regions of cells is a mélange of twisted and interwoven fine threads, the chromosomes. Chromosomes by weight are composed of 50–60 percent protein and 40–50 percent DNA. During cell division, in all cells but those of bacteria (and some ancestral protists), the chromosomes display an elegantly choreographed movement, separating so that each offspring of the original cell receives an equal complement of chromosomal material. This pattern of segregation corresponds in all details to the theoretically predicted pattern of segregation of the genetic material implied by the fundamental genetic laws. The chromosome combination of the DNA and the proteins (histone or protamine) is called nucleoprotein. The DNA stripped of its protein is known to carry genetic information and to determine details of proteins produced in the cytoplasm of cells; the proteins in nucleoprotein regulate the shape, behaviour, and activities of the chromosomes themselves.

The other major nucleic acid is ribonucleic acid (RNA). Its five-carbon sugar is slightly different from that of DNA. Thymine, one of the four bases that make up DNA, is replaced in RNA by the base uracil. RNA appears in a single-stranded form rather than a double. Proteins (including all enzymes), DNA, and RNA have a curiously interconnected relation that appears ubiquitous in all organisms on Earth today. RNA, which can replicate itself as well as code for protein, may be older than DNA in the history of life.

Modes of Nutrition and Energy Generation

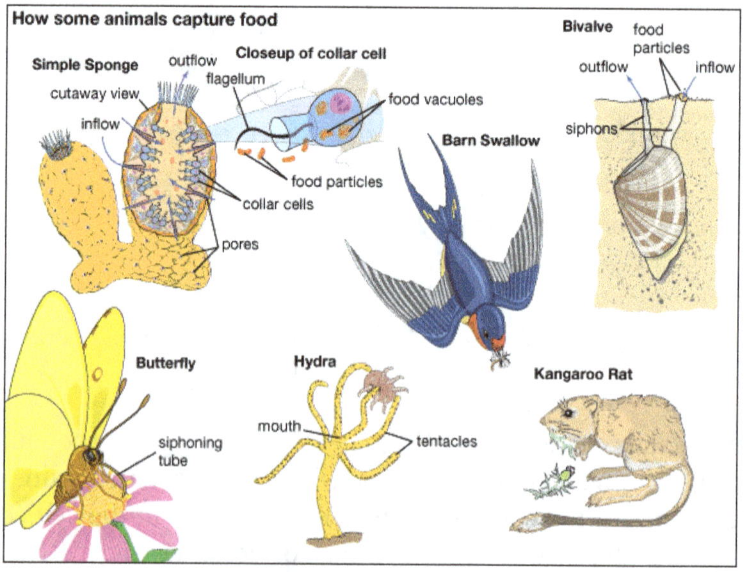

Food capture: Animals.

Chemical bonds that make up the compounds of living organisms have a certain probability of spontaneous breakage. Accordingly, mechanisms exist that repair this damage or replace the broken molecules. Furthermore, the meticulous control that cells exercise over their internal activities requires the continued synthesis of new molecules.

Processes of synthesis and breakdown of the molecular components of cells are collectively termed metabolism. For synthesis to keep ahead of the thermodynamic tendencies toward breakdown, energy must be continuously supplied to the living system.

Organisms acquire energy by two general methods: by light or by chemical oxidation. Productive organisms, called autotrophs, convert light or chemicals into energy-rich organic compounds beginning with energy-poor carbon dioxide (CO_2). These autotrophs provide energy for the other organisms, the heterotrophs. Heterotrophs are organisms that acquire their energy by the controlled breakdown of preexisting organic molecules, or food. Human beings, like most other animals, fungi, protists, and bacteria, are heterotrophs.

Autotrophic organisms are often primary producers in their ecosystems. They acquire their useful free energy from sources other than food: either from the energy of sunlight (photoautotrophs) or from oxidative chemical reactions (chemoautotrophs). The latter mode of metabolism refers to life-forms that use inorganic materials (ammonia [NH_3], methane [CH_4], or hydrogen sulfide [H_2S]) combined with oxygen to generate their energy. Only some bacteria are capable of obtaining energy by "burning" inorganic chemicals.

Green plants are typical photoautotrophs. Plants absorb sunlight to generate ATP and to disassociate water into oxygen and hydrogen. To break down the water molecule, H_2O, into hydrogen and oxygen requires much energy. The hydrogen from water is then combined in the "dark reactions" with carbon dioxide, CO_2. The result is the production of such energy-rich organic molecules as sugars, amino acids, and nucleotides. The oxygen becomes the gas O_2, which is released as waste back into the atmosphere. Animals, which are strictly heterotrophs, cannot live on carbon dioxide, sunlight, and water with a few salts like plants do. They must breathe in the atmospheric oxygen. Animals combine oxygen chemically with hydrogen atoms that they remove from their food—that is, from organic materials such as sugar, protein, and amino acids. Animals release water as a waste product from the oxygen respiration. Animals, like all heterotrophs, use organic materials as their sole source of carbon. This conversion of carbon provides an example of an aspect of an ecological cycle in which a required element flows through different types of organisms as it changes its oxidation state from CO_2 to (CH_2O_n) and back to CO_2.

Metabolic cycles in general—the extraction by organisms of useful energy and food molecules from environmental source material—can be described in terms of oxidation-reduction reactions. In the case of oxygen respiration, oxygen molecules from the air accept electrons ultimately from glucose or amino acids. The oxygen, which has a great affinity for electrons, is called an electron acceptor, whereas the glucose, or other sugar or organic molecules, is an electron donor. Animal respiration is the prototype of oxidation-reduction reactions, but certainly not all oxidation-reduction reactions (or "redox reactions," as they are often called) involve oxygen. Many other inorganic compounds are respired, or "breathed," at the cell level. Biological electron acceptors besides oxygen include nitrate, nitrite, sulfate, carbonate, elemental sulfur, and methanol. Biological electron donors (other than sugar and amino acids) include hydrogen, nitrogen compounds (as ammonia, nitrite), sulfide, and methane. For acceptor-donor transformations to be available to chemoautotrophs and heterotrophs over sustained periods of time, ecological cycles are required. For geologically short periods of time, organisms may live off a finite supply of material; however, for any long-term continuance of life, a dynamic cycling of matter involving complementary types of organisms must prevail. If life exists on other planets, the requisite elements and liquid water must cycle. A search for such transformations provides one method of detecting extraterrestrial life.

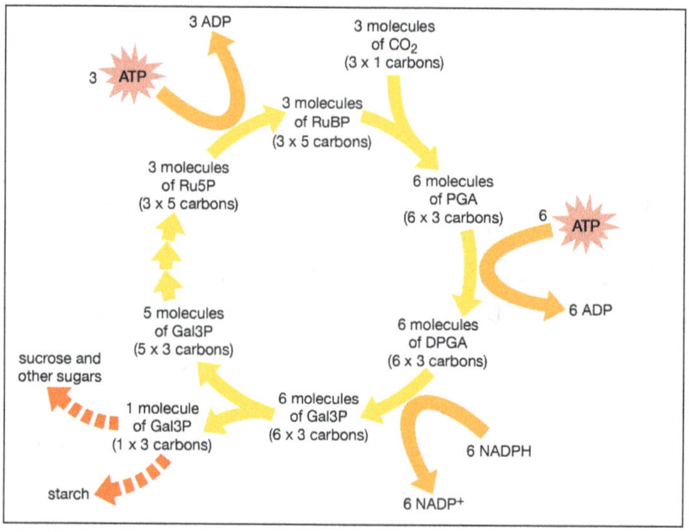

Pathway of carbon dioxide fixation and reduction in photosynthesis, the Calvin cycle.

The figure represents one complete turn of the cycle, with the net production of one molecule of Glyceraldehyde-3-phosphate (Gal3P). This three-carbon sugar phosphate usually is converted to either sucrose or starch.

In addition to energy, all forms of life require carbon sources. Autotrophic organisms (chemosynthetic and photosynthetic bacteria, algae, and plants) derive this essential element from carbon dioxide. Heterotrophs use preformed organic compounds as their source of carbon. Among autotrophs many types of cells do not depend on light to generate ATP; those that do without light are the chemoautotrophic bacteria, including the methanogens, ammonia oxidizers, sulfide oxidizers, hydrogen oxidizers, and a few obscure others. Indeed, at least five metabolic pathways entirely different from each other have evolved to use carbon dioxide gas. One is the oxygenic pathway described above, which is used by plants, algae, and cyanobacteria: the Calvin-Benson dark reactions. Other, more obscure pathways include Phosphoenolpyruvate (PEP), succinate, and methanogen pathways. They all need to bring energy-poor carbon dioxide into the energy-rich carbon-hydrogen compound metabolism of organisms. All life on Earth depends on these autotrophic reactions that begin with carbon dioxide or its equivalent. Equivalents as carbon sources in autotrophic metabolism include the carbonate ion, bicarbonate ion, and carbon monoxide. As usual, with respect to metabolic variation and virtuosity, the bacterial repertoire is vastly more diverse than that of eukaryotes—that is, plants, animals, and other organisms composed of cells with nuclei. In general, nucleated organisms, eukaryotes, are either photolithoautotrophs (i.e., algae and plants) that derive energy from light or minerals or chemo-organoheterotrophs (animals, fungi, and most protists) that derive energy and carbon from preformed organic compounds (food).

ATP

All ATP biological electron-transfer reactions lead to the net production of ATP molecules. Two of the three phosphates (PO_4) of this molecule are held by energy-rich bonds sufficiently stable to survive for long periods of time in the cell but not so strong that the cell cannot tap these bonds for energy when needed. ATP and similar molecules (such as Guanosine Triphosphate [GTP]) have a five-carbon sugar and three phosphates. As far as is known, such molecules are the general and unique energy currency of living systems on Earth.

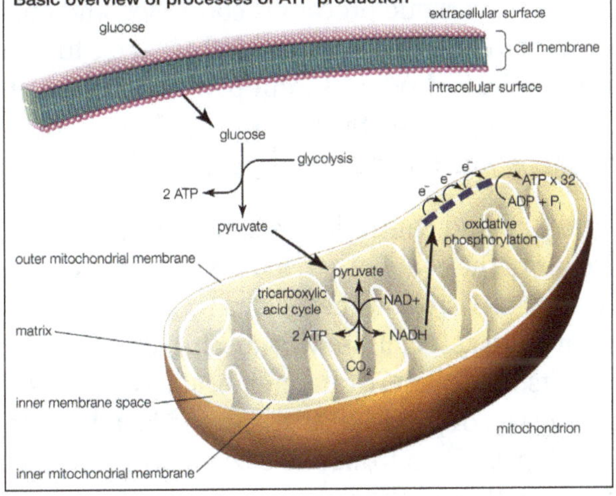

Basic overview of processes of ATP production

The three processes of ATP production include glycolysis, the tricarboxylic acid cycle, and oxidative phosphorylation. In eukaryotic cells the latter two processes occur within mitochondria. Electrons that are passed through the electron transport chain ultimately generate free energy capable of driving the phosphorylation of ADP.

No metabolic process occurs in a single step. The ordinary six-carbon sugar, glucose, does not oxidize to carbon dioxide and water in living cells in the same way that glucose in air burns. Any release of energy by burning would be too sudden and too concentrated in a small volume to happen safely inside the tiny cell. Instead, glucose is broken down at ambient (i.e., relatively cool) temperatures by a series of successive and coordinated steps. Each step is mediated by a particular and specific enzyme. In most cells that metabolize glucose, the sugar first breaks down in a set of steps that occur in the absence of oxygen. The total number of such steps in plants, animals, fungi, and protists is about 11. Other organisms, primarily bacteria and obscure protists and fungi, are anaerobes: they do not utilize molecular oxygen in their metabolism. In anaerobes, glucose metabolism stops at compounds such as ethanol or lactic acid. Aerobic organisms, including all animals, carry the oxidation of glucose farther. They rapidly use anaerobic glucose breakdown products such as lactic acid, ethanol, or acetate with Krebs-cycle intermediates in the mitochondria. Aerobic oxidation of glucose requires an additional 60 enzyme-catalyzed steps. The anaerobic breakdown of glucose uses enzymes suspended freely in solution in the cells. The aerobic steps occur on enzymes localized in mitochondria, the "power packs" of cells where oxygen gas is used to make the energy compound ATP. The complete aerobic breakdown of sugar to carbon dioxide and water is about 10 times more efficient than the anaerobic in that 10 times as many ATP molecules are produced.

Energy made available to cells in the form of ATP is used in a variety of ways—for example, for motility. When an amoeba extends pseudopods or when a person walks, ATP molecules are tapped for their energy-rich phosphate bonds. ATP molecules are used for the synthesis of proteins that all cells require in their growth and division, amino acids, and five-carbon sugars of nucleic acids. Each synthetic process is controlled and enzymatically mediated. Each starts from an organic building-block compound available to the cell as food. The amino acid L-leucine, for example, is produced from pyruvic acid, which is itself the product of the anaerobic breakdown of glucose. Synthesis of L-leucine from pyruvic acid involves eight enzyme-mediated steps with an addition of acetic acid and water.

These exquisitely interlocked and controlled metabolic steps are not performed in a diffuse manner helter-skelter in the cell. Rather, a marvelously architectured cellular interior displays specialized regions visible at the electron-microscopic level. Particular chemical reactions are performed in association with specific structures. In aerobic eukaryotes the mitochondrion with its intricate cristate membrane (the folds in the membrane are called cristae) provides the site of pyruvate, acetate, and lactate metabolism. These molecules are transformed and passed on from one enzyme to another as through a conveyor belt in a factory. Similarly, in those eukaryotes capable of oxygenic photosynthesis (algae, plants), photosynthesis occurs only in an organelle (a cell part) called a chloroplast. Chlorophyll, carotenoids, and other pigments that absorb visible light, as well as the detailed enzymatic apparatus for the photosynthetic process, reside there.

Chloroplasts and mitochondria contain DNA. Moreover, this DNA has a sequence distribution that differs entirely from that of the nucleus and greatly resembles that of free-living photosynthetic and oxygen-respiring bacteria. The best explanation for these facts is that the ancestors to the oxygen-releasing chloroplasts and oxygen-respiring mitochondria were once free-living bacteria.

Diversity

Prokaryotes and Eukaryotes

All life is composed of cells of one of two types: prokaryotes (those that lack a nucleus) or eukaryotes (those with a nucleus). Even in one-celled organisms this distinction is very clear.

All bacteria are prokaryotic, even though many, probably most, are multicelled in nature. The only other single-celled organisms that exist are fungi (one-celled fungi are called yeasts). All nucleated organisms (cells with nuclei and chromosomes in their cells) that are not animals, fungi, or plants are Protista. This huge group includes the unicellular or few-celled protists and their multicellular descendants. The large kingdom of Protista has 250,000 estimated species alive today. Some are very large, such as red algae and the kelp Macrocystis. One-celled protists include the familiar amoebas, paramecia, and euglenas as well as 50,000 less-familiar types. Scientifically speaking, no such thing as a one-celled animal exists. All animals and plants are by definition multicellular, since they all develop from multicellular embryos. Accordingly, all "protozoans" are now classified as single-celled Protista, not animals. Nor are there any one-celled plants. Organisms formerly called one-celled plants are algae and, as such, are now classified with Protista as well. If a mature organism is determined to be one-celled, then it must be either a bacterium (prokaryotic) or a fungus or protist (eukaryotic). All animals and plants develop from embryos that by definition combine two complementary sets of chromosomes (i.e., they are diploids at some stage in their development). They are all multicellular eukaryotes. But though there are no one-celled plants or animals, there are indeed myriad many-celled protists. Multicellularity evolved not only in the ancestors to the plants and the animals but also in the bacteria, the protists, and the fungi.

All eukaryotic cells undergo some form of mitosis, a sequence of cell division events that occurs after chromosomal DNA protein replication. Mitosis ensures that chromosomal DNA and protein are equally distributed to the offspring cells. Mitosis is the most distinctive activity of eukaryotic cells, which have nucleoprotein chromosomes in their nuclei and a membrane that separates the nucleus from the cytoplasm. In mitosis, mitochondria, which are usually present in the cytoplasm as well as in the chloroplasts of algae and plants, are smoothly distributed with the chromosomes to offspring cells. The Golgi apparatus and endoplasmic reticulum (ER), an intricately convoluted structure, serve to anchor many cytoplasmic enzymes excluded from mitochondria or chloroplasts. They also divide and are distributed in mitosis.

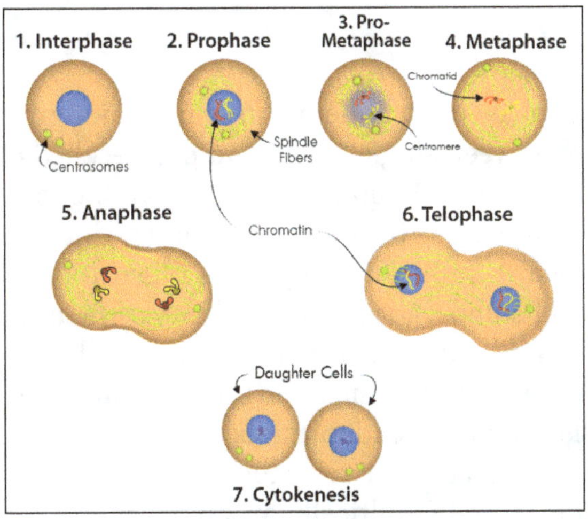

Stages of mitosis: (A) Prophase. Replicated chromosomes, consisting of two daughter strands (chromatids) attached by a centromere, coil and contract. Two pairs of specialized organelles (centrioles) begin to move apart, forming a bridge of hollow protein cylinders known as microtubules (spindle fibres) between them. Microtubules also extend in a radial array (aster) from the centrioles to the poles of the cell. (B) Late prophase. As the centrioles move apart, the nuclear membrane breaks down and microtubules extend from each centromere to opposite sides or poles of the cell. (C) Metaphase. The centromeres align in a plane midway between the poles known as the equator, or metaphase plate. During late metaphase, each centromere divides into two, freeing sister chromatids from each other. (D) Anaphase. Sister chromatids are drawn to opposite ends as centromeric microtubules shorten and polar microtubules lengthen, causing the poles to move farther apart. (E) Telophase. Chromosomes uncoil, microtubules disappear, and the nuclear membrane re-forms around each set of daughter chromosomes. The cytoplasm begins to pinch in to create two daughter cells.

Nuclei, chromosomes, mitochondria, chloroplasts, ER, and nuclear membranes are all absent in prokaryotes. Prokaryotic cells, which include all the cyanobacteria (formerly called blue-green algae), are bacteria in every way. Division is nonmitotic in all prokaryotes. Bacteria lack nucleoprotein and a nuclear membrane, and, when chromosome stain is applied, only fuzz or nothing is seen. Whereas all eukaryotic cells have more than one chromosome and sometimes over a thousand, the genes of prokaryotic cells are organized into a single "chromoneme" or "genophore." (The term bacterial chromosome, while still in use, is, technically speaking, inaccurate.) The genes may or may not be concentrated enough to be seen, but in any case bacterial DNA floats freely in the cytoplasm. Prokaryote cell organization is less complex than that of eukaryotes. The basic question of the evolution of prokaryotes into eukaryotes—often rated the second major evolutionary mystery, after the origin of life—is thought to involve a complex series of partnerships in which distinct strains of bacteria entered each others' bodies, merged symbiotically, and traded genes.

Multicellularity

Since multicellularity evolved independently in every major group of microorganisms, the blurred distinction between single-celled and many-celled organisms has become obsolete. The protists are divisible into about 35 unambiguous groups called phyla. They provide many examples of biological principles—including the prevalence of independent trends toward multicellularity. One illustration involves cellular slime molds. These heterotrophs undergo an extraordinary sequence of events during their life history. The story begins with single cells, indistinguishable from common amoebas. When starved, they begin to swarm. Soon they combine into a slimy mass of many nucleated amoeba cells called a pseudoplasmodium. The pseudoplasmodium in turn forms a sluglike multicellular creature resembling a mollusk that has escaped from its shell. This slug, which is entirely multicellular, migrates and then stops and develops into a stalk structure called a sorocarp that bears amoeba cysts on top. The cysts were called "spores." Some have cellulose cell walls similar to those of plants. The cysts, which are encased amoebas (just like other amoeba cysts), germinate in turn—when water and food again become abundant—into new amoebas. The released amoebas extend their pseudopods, and, as individuals again, they migrate to feed. The life history repeats with swarms of migrating amoebas, slugs, stalks, and finally clusters of amoeba cysts on top as wet, food-rich conditions are followed by dryness and scarcity.

Biology is replete with life histories of comparable or even greater complexity. In protist life histories—by far the most diverse, exotic, and unique—one can search for ancestral modes of life, including missing links between the prokaryotes and eukaryotes. The one-celled swimming stage is called a sperm, whose imperative it is to find another one-celled partner, the ovum. Like all animals, humans develop from a single fertilized egg with its complement of two sets of genetic material. These diploid fertilized egg cells then divide to form many presumably identical cells. The early embryology of all animals goes through stages that have 2, 4, 8, 16, and so on cells. Genetic information is theoretically identical in each cell. But then how does it ever happen that, as they mature, the cells become permanently specialized to form hair, bone, liver, blood, or nerve cells? How does any given cell "know" what sort of specialized cell it must become, since all cells seem to contain identical nucleic acids? Despite a century of work on this process (called differentiation) and the discovery of many facts about embryos, this basic problem still remains unsolved in animal biology.

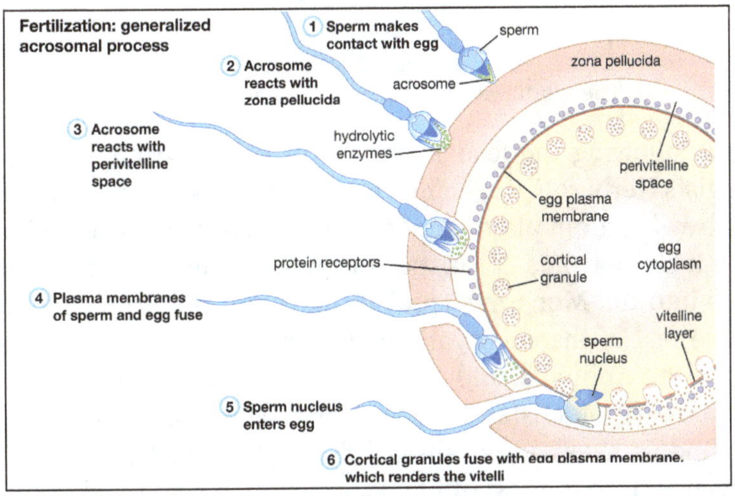

Fertilization of a human egg.

(1) The sperm release enzymes that help disperse the corona radiata and bind to the zona pellucida. (2) The outer sperm head layer is sloughed off, exposing enzymes that digest a path through the zona pellucida. (3) The sperm fuses with the egg cell membrane, causing the zona pellucida to become impenetrable to other sperm. (4) The tail separates from the sperm head, and the male pronucleus enlarges and travels to the female pronucleus in the center of the cell. Chromosomes merge to form a fertilized egg.

Classification and Microbiota

Prior to the recognition of microbial life, the living world was too easily divided into animals that moved in pursuit of food and plants that produced food from sunlight. The futility of this simplistic classification scheme has been underscored by entire fields of science. Many bacteria both swim (like animals) and photosynthesize (like plants), yet they are best considered neither. Many algae (e.g., euglenids, dinomastigotes, chlorophytes) also swim and photosynthesize simultaneously. Molecular biological measurements of the DNA that codes for components of the ribosomes (organelles that are universally distributed) consistently find fungi to be extremely different from plants. Indeed, fungi genetically resemble animals more than plants.

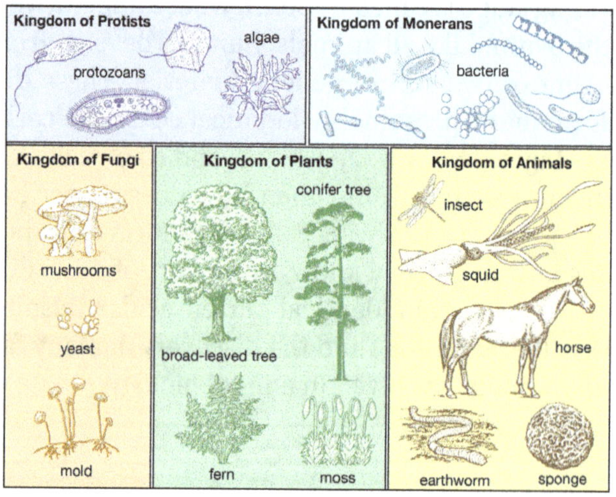

The Whittaker five-kingdom classification of life.

Modern biology, following the lead of the German biologist Ernst Haeckel and the American biologists Herbert F. Copeland and Robert H. Whittaker, has now thoroughly abandoned the two-kingdom plant-versus-animal dichotomy. Haeckel proposed three kingdoms when he established "Protista" for microorganisms. Copeland classified the microorganisms into the Monerans (prokaryotes) and the Protoctista (which included fungi with the rest of the eukaryotic microorganisms). His four-kingdom scheme (Monera, Protoctista, Animalia, and Plantae) had the advantage of clearly separating microbes with nuclei (Protoctista) from those without (Monera: the prokaryotes—that is, the bacteria and archaea) and of distinguishing the two embryo-forming groups—plants and animals—from the rest of life. Whittaker, on ecological grounds, raised the fungi to kingdom status. The modified Whittaker five-kingdom classification system is perhaps the most comprehensible and biologically based way to unambiguously organize information about all groups of living beings. American microbiologist Carl Woese has offered still another classification scheme, in which all organisms are placed in either the Archaea (prokaryotes that include some salt lovers, acid lovers, and methane producers), the Bacteria (all other prokaryotes, including obligate anaerobic bacteria as well as photosynthetic and chemoautotrophic bacteria), or the Eukarya (all eukaryote forms of life). Woese's scheme is based on molecular biological criteria that focus on the RNA sequence of morphological factors to classify new or disputed organisms. Although Woese's three-domain system is very popular, a potential problem with it is that RNA, one characteristic among thousands, does not consistently correlate with many others.

Microbes (or microbiota) are simply all those organisms too small to be visualized without some sort of microscopy. Bacteria, the smaller fungi, and the smaller protists are undoubtedly microbes. Some scientists classify tiny animals, worms, and rotifers as microbes as well. Like weed, a plant not wanted in a garden, microbe is often a more useful term than one with a precise scientific meaning.

Sizes of Organisms

The sizes of organisms on Earth vary greatly and are not always easy to estimate. On the large end, great stands of poplar trees entirely connected by common roots are really a single organism. A variety of influences place an upper limit to the size of organisms. One is the strength of biological materials. Sequoia redwood trees, some of which exceed 90 metres (300 feet), are apparently near the upper limit of height for an organism. The Italian astronomer Galileo calculated in 1638 that a tree taller than roughly 90 metres would buckle under its own weight when displaced slightly from the vertical (for example, by a breeze). Because of the buoyancy of water, large animals such as whales are not presented with such stability problems. Other size-related difficulties arise. The volume of tissues to be nourished increases as the cube of the characteristic length of the organism, but the surface of the gut, which absorbs the ingested food, increases only as the square of the length for a fixed morphology. As an organism's length is increased, a point of diminishing returns is ultimately reached where nutrition is irreversibly impeded in an animal.

New work on genome sequences, the total amount and quality of all of the genes that make up a live being, permits more accurate assessment of the material basis of the theoretically smallest and simplest extant free-living organisms. The complete DNA sequences of a few extremely small free-living organisms are now known—e.g., Mycoplasma genitalium with its 480 genes. All the molecules necessary for metabolism must be present. The smallest free-living cells include the pleuropneumonia-like organisms (PPLOs). Whereas an amoeba has a mass of 5×10^{-7} gram (2×10^{-8} ounce), a PPLO, which cannot be seen without a high-powered electron microscope, weighs 5×10^{-16} gram (2×10^{-15} ounce) and is only about 100 nanometres across. PPLOs grow very slowly. Other, even smaller organisms that grow even more slowly would be extremely difficult to detect. An organism the size of a PPLO that has room for only about a hundred enzymes depends entirely upon the animal tissue in which it lives. A much smaller organism would have room for many fewer enzymes. Its ability to accomplish the functions required for autopoiesis in living systems would be severely compromised. Were there, however, an environment in which all the necessary organic building blocks and such energy sources as ATP were provided "free," then there might be a functioning organism substantially smaller than a PPLO. The inside of cells provides just such an environment, which explains why infectious agents, such as prions, plasmids, and viruses, may be substantially smaller than PPLOs. But it must be emphasized that viruses and their kin are not, even in principle, autopoietic.

Metabolites and Water

The range of organic molecules that organisms, especially microbes, can metabolize is very wide and occasionally includes foods such as formaldehyde or petroleum that seem unlikely from a human point of view. Pseudomonas bacteria are capable of using almost any organic molecule as a source of carbon and energy, provided only that the molecule is at least slightly soluble in water. Microorganisms cannot metabolize plastics, not because of any fundamental chemical prohibitions but probably because plastics have not been part of the environment of microorganisms for very long. A lack of oxygen is thought of as extremely deleterious to life, but this view is anthropocentric. Many bacteria are facultative anaerobes that can take their oxygen or leave it. Many other bacteria and protists are obligate anaerobes that are actually poisoned by oxygen.

Water, which is crucial for life, is the major molecule in all organisms. Unless a massive mineral skeleton is present, the dry matter of most organisms is about one-half carbon by weight. This reflects the fact that all organic molecules are composed of carbon bound at least to hydrogen. Metabolism uses a wide variety of other chemical elements. Amino acids are made of nitrogen and sulfur in addition to carbon, hydrogen, and oxygen. Nucleic acids are made of phosphorus in addition to hydrogen, nitrogen, oxygen, and carbon. Sodium, potassium, and calcium are used to maintain electrolyte balance and to signal cells. Silicon is used as a structural material in the diatom shell, the radiolarian and heliozoan spicule, and the chrysophyte exoskeleton. Iron plays a fundamental role in the transport of molecular oxygen as part of the hemoglobin molecule. In some ascidians (sea squirts), however, vanadium replaces iron. Ascidian blood also contains unusually large amounts of niobium, titanium, chromium, manganese, molybdenum, and tungsten. The vanadium and niobium compounds in ascidian blood may be adaptations to low oxygen levels. Some bacteria use selenium, tellurium, or even arsenic as electron acceptors. Others produce the fully saturated gas hydrides of carbon, arsenic, phosphorus, or silicon as a metabolic waste. Still others form compounds of carbon with such halogens as chlorine or iodine. Not only the foregoing elements but also copper, zinc, cobalt, and possibly gallium, boron, and scandium perform particular functions in the enzymatic apparatus of particular cells. These elements, both the uncommon ones and those as common as phosphorus, are much more concentrated in living matter than in the environment where the living matter resides. This concentration suggests that such rare chemicals play unique functional roles that other, more abundant elements cannot serve.

Sensory Capabilities and Awareness

Although any given organism is severely limited in its range of behaviour patterns and sensory capabilities, life as a whole is remarkably sensitive to aspects of its local social and physical environment. A bird raised from the egg in the absence of other members of its species migrates when the season beckons, builds the proper nest, and engages in elaborate courtship rituals. Those birds that fail to perpetuate the behaviour pattern do not leave descendants. Such behavioural accuracy itself must have evolved. Rats that pass through mazes easily interbreed, as do rats that pass through with difficulty; eventually two populations with inherited characteristics called "maze-smart" and "maze-dumb" are produced. Fruit fly populations attracted to light can be separated from those that avoid light. Classical genetic-crossing experiments reveal that the two populations differ largely in a small number of genes for phototropism. Similar genetic determinants of behaviour exist in humans. For example, possession of a supernumerary Y-chromosome in males is strikingly correlated with aggressive tendencies.

Arctic tern (Sterna paradisaea). It breeds in the southerly reaches of the Arctic and winters in the Antarctic regions, thus making the longest annual migration of any bird.

Photosensitivity, audiosensitivity, thermosensitivity, chemosensitivity, and magnetosensitivity: Humans use only a limited region of the electromagnetic spectrum, the part called visible light, which extends from 400 to 700 nanometres in wavelength. While plants, algae, photosynthetic bacteria, and most animals are sensitive to this same range of wavelengths, many are sensitive to other wavelengths as well. Many plants present flower patterns visible only in the ultraviolet range at wavelengths below 400 nanometres, where pollinating insects are sensitive. Honeybees use polarized light—which the unaided human eye is unable to detect—for direction finding on partly cloudy days. The "pit" of such pit vipers as the rattlesnake is an infrared (heat) receptor that serves as a direction finder. These reptiles sense the thermal radiation emitted by mammals and birds, their warm-blooded prey. Humans are entirely insensitive to this thermal radiation.

An evening primrose (Oenothera biennis) seen (top) in visible light and (bottom) in ultraviolet light; the latter reveals nectar-guide patterns that are discernible to the moth pollinating this flower but not to the human eye.

That some animals such as dogs are sensitive to sounds that the human ear cannot detect is obvious to those who use dog whistles. Bats emit and detect sound waves at ultrahigh frequencies, in the vicinity of 100,000 cycles per second, about five times the highest frequency to which the human ear is sensitive. Bats have echolocated their prey by use of these sounds for millions of years before humans invented radar and sonar. The audio receptors of many moths that are prey to bats respond only to the frequencies emitted by the bats. When the bat sounds are heard, the moths take evasive action. Dolphins communicate via a very wide frequency range. They employ a "click" echolocator.

Epauletted fruit bat (Epomophorus wahlbergi).

Some species of animals enjoy highly specialized and exotic organs for the detection or transmission of sound. Dolphins and whales use their blowholes rather than their mouths to utter their sounds.

Smell and taste, or some form of detection of specific chemical molecules, are universal. The ultimate in olfactory specialization may be male moths, whose feathery antennae are underlain by splayed microtubules, each of which is covered by a membrane at the distal end. They smell essentially nothing except the epoxide compound called disparlure, the chemical sex attractant discharged by the female. Only 40 molecules per second need impact on the antennae to produce a marked response. One female silkworm moth need release only 10^{-8} gram (4×10^{-10} ounce) of sex attractant per second in order to attract every male silkworm moth within a few kilometres.

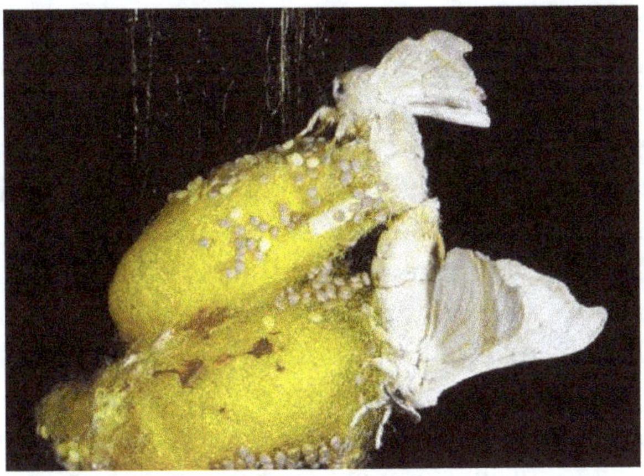

Silkworm moth.

Magnetotactic bacteria sense Earth's magnetic field. North Pole-seeking bacteria swim toward the sediment-water interface as they follow the magnetic lines of force. South Pole-seeking flagellated magnetotactic bacteria do the same in the Southern Hemisphere. Since those studied are microaerophiles—i.e., they require oxygen in lower than ambient concentrations—pole seekers tend to arrive at oxygen-depleted sediment adequate for their continued growth and reproduction. Ultrastructural studies reveal magnetosomes, tiny single-domain crystals of magnetite, an iron oxide mineral sensitive to magnetic fields, or greigite, an iron sulfide mineral, in their cells. The magnetosomes are aligned along the axis of the cell and serve to orient the sensitive bacteria. All the different kinds of magnetotactic bacteria bear magnetosomes in their cells. Whether magnetotaxis is causal in the orientation of homing pigeons, dancing bees on cloudy days, or other instinctively orienting animals is under investigation.

Besides the familiar senses of sight, hearing, smell, taste, and touch, organisms have a wide variety of other senses. People have inertial orientation systems and accelerometers in the cochlear canal of the ear. The water scorpion (Nepa) has a fathometer sensitive to hydrostatic pressure gradients. Many plants have chemically amplified gravity sensors made of modified chloroplasts. Some green algae use barium sulfate and calcium ion detection systems to sense gravity. Fireflies and squids communicate with their own kind by producing changing patterns of light on their bodies. The nocturnal African freshwater fish Gymnarchus niloticus operates a dipole electrostatic field generator and a sensor to detect the amplitude and frequency of disturbances in turbulent waters.

Sensing with Technology

From the foregoing sample of sensibilities, it is clear that a vast repertoire of sensors in living beings confers upon their possessors an awareness of the environment that differs from humanity's. Humans, however, have an enhanced ability to extend their sensory and intellectual capabilities through the use of instrumentation far beyond those with which they are born.

The sensory system of Earth is expanded by the capabilities of human machines. From those that detect ionizing radiation, wind velocities, the taste of wine, the concentration of salt in solution, a few photons of light in a dark corridor, or the blood temperature of an infant to those that record microearthquakes, a lying smile, or the heat of a furnace, the sensory systems of the biosphere—nonhuman, human, and human-mediated—have augmented over time. Indeed, just as seeing-eye dogs transmit visual information to their blind owners, the sensory system of life extends far beyond any given species of animals and its machines to the entire sensitive biota in this pulsating biosphere. Sensitivity to sound, chemicals, heat, light, mechanical movement, magnetism, and charged particles has been tallied by many a hardworking scientist. Whether entire categories of sensory information are missing from that list is not entirely clear. Great sensitivity to the environment abounds even in those smallest life-forms, the bacteria. Life has been sensing and responding to its environment since its inception more than 3.7 billion to 3.5 billion years ago. Moreover, it is not clear at what point in evolutionary history, or where precisely among organisms, consciousness comes in. Humans are conscious and self-conscious. But are protists that choose certain shapes and sizes of glass beads over others conscious of their decision making? Charles Darwin recognized selection among various male suitors by females as instrumental in the evolution of sexual species, including birds and insects. The extent to which consciousness and choice making are important in evolution remains a matter for debate.

AGE OF THE EARTH

The age of the Earth is estimated to be 4.54 ± 0.05 billion years (4.54×10^9 years $\pm$ 1%). This age may represent the age of the Earth's accretion, of core formation, or of the material from which the Earth formed. This dating is based on evidence from radiometric age-dating of meteorite material and is consistent with the radiometric ages of the oldest-known terrestrial and lunar samples.

Following the development of radiometric age-dating in the early 20th century, measurements of lead in uranium-rich minerals showed that some were in excess of a billion years old. The oldest such minerals analyzed to date, small crystals of zircon from the Jack Hills of Western Australia are at least 4.404 billion years old. Calcium–aluminium-rich inclusions, the oldest known solid constituents within meteorites that are formed within the Solar system are 4.567 billion years old, giving a lower limit for the age of the Solar System.

It is hypothesised that the accretion of Earth began soon after the formation of the calcium-aluminium-rich inclusions and the meteorites. Because the time this accretion process took is not yet known, and predictions from different accretion models range from a few million up to about 100 million years, the difference between the age of Earth and of the oldest rocks is difficult to determine. It is also difficult to determine the exact age of the oldest rocks on Earth, exposed at the surface, as they are aggregates of minerals of possibly different ages.

Development of Modern Geologic Concepts

Studies of strata, the layering of rocks and earth, gave naturalists an appreciation that Earth may have been through many changes during its existence. These layers often contained fossilized remains of unknown creatures, leading some to interpret a progression of organisms from layer to layer.

Nicolas Steno in the 17th century was one of the first naturalists to appreciate the connection between fossil remains and strata. His observations led him to formulate important stratigraphic concepts (i.e., the "law of superposition" and the "principle of original horizontality"). In the 1790s, William Smith hypothesized that if two layers of rock at widely differing locations contained similar fossils, then it was very plausible that the layers were the same age. William Smith's nephew and student, John Phillips, later calculated by such means that Earth was about 96 million years old.

In the mid-18th century, the naturalist Mikhail Lomonosov suggested that Earth had been created separately from, and several hundred thousand years before, the rest of the universe. Lomonosov's ideas were mostly speculative. In 1779 the Comte du Buffon tried to obtain a value for the age of Earth using an experiment: He created a small globe that resembled Earth in composition and then measured its rate of cooling. This led him to estimate that Earth was about 75,000 years old.

Other naturalists used these hypotheses to construct a history of Earth, though their timelines were inexact as they did not know how long it took to lay down stratigraphic layers. In 1830, geologist Charles Lyell, developing ideas found in James Hutton's works, popularized the concept that the features of Earth were in perpetual change, eroding and reforming continuously, and the rate of this change was roughly constant. This was a challenge to the traditional view, which saw the history of Earth as static, with changes brought about by intermittent catastrophes. Many naturalists were influenced by Lyell to become "uniformitarians" who believed that changes were constant and uniform.

Early Calculations

In 1862, the physicist William Thomson, 1st Baron Kelvin published calculations that fixed the age of Earth at between 20 million and 400 million years. He assumed that Earth had formed as a completely molten object, and determined the amount of time it would take for the near-surface temperature gradient to decrease to its present value. His calculations did not account for heat produced via radioactive decay (a then unknown process) or, more significantly, convection inside the Earth, which allows the temperature in the upper mantle to remain high much longer, maintaining a high thermal gradient in the crust much longer. Even more constraining were Kelvin's estimates of the age of the Sun, which were based on estimates of its thermal output and a theory that the Sun obtains its energy from gravitational collapse; Kelvin estimated that the Sun is about 20 million years old.

William Thomson (Lord Kelvin)

Geologists such as Charles Lyell had trouble accepting such a short age for Earth. For biologists, even 100 million years seemed much too short to be plausible. In Darwin's theory of evolution, the process of random heritable variation with cumulative selection requires great durations of time. According to modern biology, the total evolutionary history from the beginning of life to today has taken place since 3.5 to 3.8 billion years ago, the amount of time which passed since the last universal ancestor of all living organisms as shown by geological dating.

In 1869, Darwin's great advocate, Thomas H. Huxley, attacked Thomson's calculations, suggesting they appeared precise in themselves but were based on faulty assumptions. The physicist Hermann von Helmholtz (in 1856) and astronomer Simon Newcomb (in 1892) contributed their own calculations of 22 and 18 million years respectively to the debate: they independently calculated the amount of time it would take for the Sun to condense down to its current diameter and brightness from the nebula of gas and dust from which it was born. Their values were consistent with Thomson's calculations. However, they assumed that the Sun was only glowing from the heat of its gravitational contraction. The process of solar nuclear fusion was not yet known to science.

In 1895 John Perry challenged Kelvin's figure on the basis of his assumptions on conductivity, and Oliver Heaviside entered the dialogue, considering it "a vehicle to display the ability of his operator method to solve problems of astonishing complexity."

Other scientists backed up Thomson's figures. Charles Darwin's son, the astronomer George H. Darwin, proposed that Earth and Moon had broken apart in their early days when they were both molten. He calculated the amount of time it would have taken for tidal friction to give Earth its current 24-hour day. His value of 56 million years added additional evidence that Thomson was on the right track.

The last estimate Thomson gave, in 1897, was: "that it was more than 20 and less than 40 million year old, and probably much nearer 20 than 40". In 1899 and 1900, John Joly calculated the rate at which the oceans should have accumulated salt from erosion processes, and determined that the oceans were about 80 to 100 million years old.

Radiometric Dating

By their chemical nature, rock minerals contain certain elements and not others; but in rocks containing radioactive isotopes, the process of radioactive decay generates exotic elements over time. By measuring the concentration of the stable end product of the decay, coupled with knowledge of the half life and initial concentration of the decaying element, the age of the rock can be calculated. Typical radioactive end products are argon from decay of potassium-40, and lead from decay of uranium and thorium. If the rock becomes molten, as happens in Earth's mantle, such nonradioactive end products typically escape or are redistributed. Thus the age of the oldest terrestrial rock gives a minimum for the age of Earth, assuming that no rock has been intact for longer than the Earth itself.

Convective Mantle and Radioactivity

In 1892, Thomson had been made Lord Kelvin in appreciation of his many scientific accomplishments. Kelvin calculated the age of the Earth by using thermal gradients, and he arrived at an estimate of about 100 million years. He did not realize that the Earth mantle was convecting, and this invalidated his estimate. In 1895, John Perry produced an age-of-Earth estimate of 2 to 3 billion years using a model of a convective mantle and thin crust, however his work was largely ignored. Kelvin stuck by his estimate of 100 million years, and later reduced it to about 20 million years.

The discovery of radioactivity introduced another factor in the calculation. After Henri Becquerel's initial discovery in 1896, Marie and Pierre Curie discovered the radioactive elements polonium and radium in 1898; and in 1903, Pierre Curie and Albert Laborde announced that radium produces enough heat to melt its own weight in ice in less than an hour. Geologists quickly realized that this upset the assumptions underlying most calculations of the age of Earth. These had assumed that the original heat of the Earth and Sun had dissipated steadily into space, but radioactive decay meant that this heat had been continually replenished. George Darwin and John Joly were the first to point this out, in 1903.

Invention of Radiometric Dating

Radioactivity, which had overthrown the old calculations, yielded a bonus by providing a basis for new calculations, in the form of radiometric dating.

Ernest Rutherford

Ernest Rutherford and Frederick Soddy jointly had continued their work on radioactive materials and concluded that radioactivity was due to a spontaneous transmutation of atomic elements. In radioactive decay, an element breaks down into another, lighter element, releasing alpha, beta, or gamma radiation in the process. They also determined that a particular isotope of a radioactive element decays into another element at a distinctive rate. This rate is given in terms of a "half-life", or the amount of time it takes half of a mass of that radioactive material to break down into its "decay product".

Some radioactive materials have short half-lives; some have long half-lives. Uranium and thorium have long half-lives, and so persist in Earth's crust, but radioactive elements with short half-lives have generally disappeared. This suggested that it might be possible to measure the age of Earth by determining the relative proportions of radioactive materials in geological samples. In reality, radioactive elements do not always decay into nonradioactive ("stable") elements directly, instead, decaying into other radioactive elements that have their own half-lives and so on, until they reach a stable element. These "decay chains", such as the uranium-radium and thorium series, were known within a few years of the discovery of radioactivity and provided a basis for constructing techniques of radiometric dating.

The pioneers of radioactivity were chemist Bertram B. Boltwood and the energetic Rutherford. Boltwood had conducted studies of radioactive materials as a consultant, and when Rutherford said at Yale in 1904, Boltwood was inspired to describe the relationships between elements in various decay series. Late in 1904, Rutherford took the first step toward radiometric dating by suggesting that the alpha particles released by radioactive decay could be trapped in a rocky material as helium atoms. At the time, Rutherford was only guessing at the relationship between alpha particles and helium atoms, but he would prove the connection four years later.

Soddy and Sir William Ramsay had just determined the rate at which radium produces alpha particles, and Rutherford proposed that he could determine the age of a rock sample by measuring its concentration of helium. He dated a rock in his possession to an age of 40 million years by this technique. Rutherford wrote, 'I came into the room, which was half dark, and presently spotted Lord Kelvin in the audience and realized that I was in trouble at the last part of my speech dealing with the age of the Earth, where my views conflicted with his. To my relief, Kelvin fell fast asleep, but as I came to the important point, I saw the old bird sit up, open an eye, and cock a baleful glance at me! Then a sudden inspiration came, and I said, "Lord Kelvin had limited the age of the Earth, provided no new source was discovered. That prophetic utterance refers to what we are now considering tonight, radium!" Behold! the old boy beamed upon me.'

Rutherford assumed that the rate of decay of radium as determined by Ramsay and Soddy was accurate, and that helium did not escape from the sample over time. Rutherford's scheme was inaccurate, but it was a useful first step.

Boltwood focused on the end products of decay series. In 1905, he suggested that lead was the final stable product of the decay of radium. It was already known that radium was an intermediate product of the decay of uranium. Rutherford joined in, outlining a decay process in which radium emitted five alpha particles through various intermediate products to end up with lead, and speculated that the radium-lead decay chain could be used to date rock samples. Boltwood did the legwork, and by the end of 1905 had provided dates for 26 separate rock samples, ranging from 92 to 570 million years. He did not publish these results, which was fortunate because they were flawed by measurement errors and poor estimates of the half-life of radium. Boltwood refined his work and finally published the results in 1907.

Boltwood's paper pointed out that samples taken from comparable layers of strata had similar lead-to-uranium ratios, and that samples from older layers had a higher proportion of lead, except where there was evidence that lead had leached out of the sample. His studies were flawed by the fact that the decay series of thorium was not understood, which led to incorrect results for samples that contained both uranium and thorium. However, his calculations were far more accurate than any that had been performed to that time. Refinements in the technique would later give ages for Boltwood's 26 samples of 410 million to 2.2 billion years.

Arthur Holmes Establishes Radiometric Dating

Although Boltwood published his paper in a prominent geological journal, the geological community had little interest in radioactivity. Boltwood gave up work on radiometric dating and went on to investigate other decay series. Rutherford remained mildly curious about the issue of the age of Earth but did little work on it.

Robert Strutt tinkered with Rutherford's helium method until 1910 and then ceased. However, Strutt's student Arthur Holmes became interested in radiometric dating and continued to work on it after everyone else had given up. Holmes focused on lead dating, because he regarded the helium method as unpromising. He performed measurements on rock samples and concluded in 1911 that the oldest (a sample from Ceylon) was about 1.6 billion years old. These calculations were not particularly trustworthy. For example, he assumed that the samples had contained only uranium and no lead when they were formed.

More important research was published in 1913. It showed that elements generally exist in multiple variants with different masses, or "isotopes". In the 1930s, isotopes would be shown to have nuclei with differing numbers of the neutral particles known as "neutrons". In that same year, other research was published establishing the rules for radioactive decay, allowing more precise identification of decay series.

Many geologists felt these new discoveries made radiometric dating so complicated as to be worthless. Holmes felt that they gave him tools to improve his techniques, and he plodded ahead with his research, publishing before and after the First World War. His work was generally ignored until the 1920s, though in 1917 Joseph Barrell, a professor of geology at Yale, redrew geological history as it was understood at the time to conform to Holmes's findings in radiometric dating. Barrell's research determined that the layers of strata had not all been laid down at the same rate, and so current rates of geological change could not be used to provide accurate timelines of the history of Earth.

Holmes' persistence finally began to pay off in 1921, when the speakers at the yearly meeting of the British Association for the Advancement of Science came to a rough consensus that Earth was a few billion years old, and that radiometric dating was credible. No great push to embrace radiometric dating followed, however, and the die-hards in the geological community stubbornly resisted. They had never cared for attempts by physicists to intrude in their domain, and had successfully ignored them so far. The growing weight of evidence finally tilted the balance in 1931, when the National Research Council of the US National Academy of Sciences decided to resolve the question of the age of Earth by appointing a committee to investigate. Holmes, being one of the few people on Earth who was trained in radiometric dating techniques, was a committee member, and in fact wrote most of the final report.

Thus, Arthur Holmes' report concluded that radioactive dating was the only reliable means of pinning down geological time scales. It described the methods used, the care with which measurements were made, and their error bars and limitations.

Modern Radiometric Dating

Radiometric dating continues to be the predominant way scientists date geologic timescales. Techniques for radioactive dating have been tested and fine-tuned on an ongoing basis since the 1960s. Forty or so different dating techniques have been utilized to date, working on a wide variety of materials. Dates for the same sample using these different techniques are in very close agreement on the age of the material.

Possible contamination problems do exist, but they have been studied and dealt with by careful investigation, leading to sample preparation procedures being minimized to limit the chance of contamination.

Why Meteorites were Used

An age of 4.55 ± 0.07 billion years, very close to today's accepted age, was determined by Clair Cameron Patterson using uranium-lead isotope dating (specifically lead-lead dating) on several meteorites including the Canyon Diablo meteorite and published in 1956.

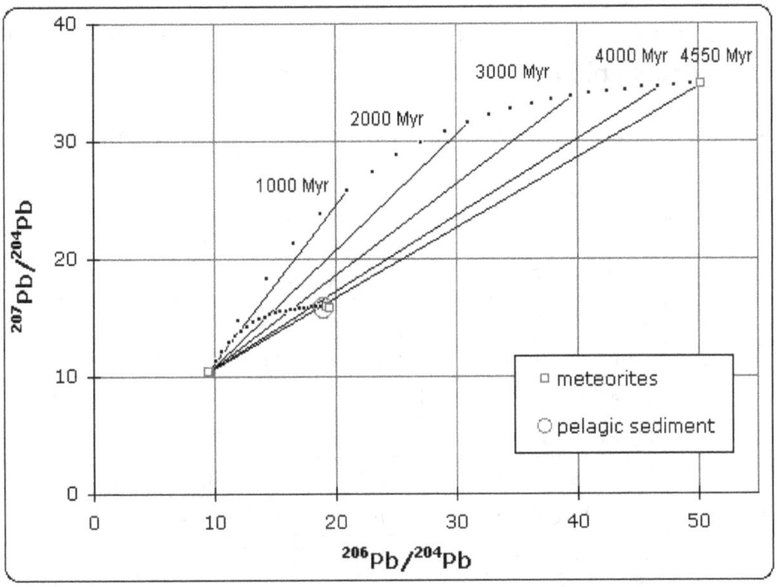

Lead isotope isochron diagram showing data used by Patterson to determine the age of the Earth in 1956.

The quoted age of Earth is derived, in part, from the Canyon Diablo meteorite for several important reasons and is built upon a modern understanding of cosmochemistry built up over decades of research.

Most geological samples from Earth are unable to give a direct date of the formation of Earth from the solar nebula because Earth has undergone differentiation into the core, mantle, and crust, and this has then undergone a long history of mixing and unmixing of these sample reservoirs by plate tectonics, weathering and hydrothermal circulation.

All of these processes may adversely affect isotopic dating mechanisms because the sample cannot always be assumed to have remained as a closed system, by which it is meant that either the parent or daughter nuclide (a species of atom characterised by the number of neutrons and protons an atom contains) or an intermediate daughter nuclide may have been partially removed from the sample, which will skew the resulting isotopic date. To mitigate this effect it is usual to date several minerals in the same sample, to provide an isochron. Alternatively, more than one dating system may be used on a sample to check the date.

Some meteorites are furthermore considered to represent the primitive material from which the accreting solar disk was formed. Some have behaved as closed systems (for some isotopic systems) soon after the solar disk and the planets formed. To date, these assumptions are supported by much scientific observation and repeated isotopic dates, and it is certainly a more robust hypothesis than that which assumes a terrestrial rock has retained its original composition.

Nevertheless, ancient Archaean lead ores of galena have been used to date the formation of Earth as these represent the earliest formed lead-only minerals on the planet and record the earliest homogeneous lead-lead isotope systems on the planet. These have returned age dates of 4.54 billion years with a precision of as little as 1% margin for error.

Statistics for several meteorites that have undergone isochron dating are as follows:

1. St. Severin (ordinary chondrite)		
1.	Pb-Pb isochron	4.543 ± 0.019 billion years
2.	Sm-Nd isochron	4.55 ± 0.33 billion years
3.	Rb-Sr isochron	4.51 ± 0.15 billion years
4.	Re-Os isochron	4.68 ± 0.15 billion years
2. Juvinas (basaltic achondrite)		
1.	Pb-Pb isochron	4.556 ± 0.012 billion years
2.	Pb-Pb isochron	4.540 ± 0.001 billion years
3.	Sm-Nd isochron	4.56 ± 0.08 billion years
4.	Rb-Sr isochron	4.50 ± 0.07 billion years
3. Allende (carbonaceous chondrite)		
1.	Pb-Pb isochron	4.553 ± 0.004 billion years
2.	Ar-Ar age spectrum	4.52 ± 0.02 billion years
3.	Ar-Ar age spectrum	4.55 ± 0.03 billion years
4.	Ar-Ar age spectrum	4.56 ± 0.05 billion years

Canyon Diablo Meteorite

Barringer Crater, Arizona where the Canyon Diablo meteorite was found.

The Canyon Diablo meteorite was used because it is both large and representative of a particularly rare type of meteorite that contains sulfide minerals (particularly troilite, FeS), metallic nickel-iron alloys, plus silicate minerals. This is important because the presence of the three mineral phases allows investigation of isotopic dates using samples that provide a great separation in concentrations between parent and daughter nuclides. This is particularly true of uranium and lead. Lead is strongly chalcophilic and is found in the sulfide at a much greater concentration than in the silicate, versus uranium. Because of this segregation in the parent and daughter nuclides during the formation of the meteorite, this allowed a much more precise date of the formation of the solar disk and hence the planets than ever before.

Fragment of the Canyon Diablo iron meteorite.

The age determined from the Canyon Diablo meteorite has been confirmed by hundreds of other age determinations, from both terrestrial samples and other meteorites. The meteorite samples, however, show a spread from 4.53 to 4.58 billion years ago. This is interpreted as the duration of formation of the solar nebula and its collapse into the solar disk to form the Sun and the planets. This 50 million year time span allows for accretion of the planets from the original solar dust and meteorites.

The Moon, as another extraterrestrial body that has not undergone plate tectonics and that has no atmosphere, provides quite precise age dates from the samples returned from the Apollo missions. Rocks returned from the Moon have been dated at a maximum of 4.51 billion years old. Martian meteorites that have landed upon Earth have also been dated to around 4.5 billion years old by lead-lead dating. Lunar samples, since they have not been disturbed by weathering, plate tectonics or material moved by organisms, can also provide dating by direct electron microscope examination of cosmic ray tracks. The accumulation of dislocations generated by high energy cosmic ray particle impacts provides another confirmation of the isotopic dates. Cosmic ray dating is only useful on material that has not been melted, since melting erases the crystalline structure of the material, and wipes away the tracks left by the particles.

Altogether, the concordance of age dates of both the earliest terrestrial lead reservoirs and all other reservoirs within the Solar System found to date are used to support the fact that Earth and the rest of the Solar System formed at around 4.53 to 4.58 billion years ago.

ORIGIN OF WATER ON EARTH

The origin of water on Earth is the subject of a significant body of research in the fields of planetary science, astronomy, and astrobiology. Earth is unique among the rocky planets in the Solar System in that it is the only planet with oceans of liquid water on its surface. Liquid water, which is necessary for life, continues to exist on the surface of Earth because the planet is distant enough from the Sun that it does not lose its water to the runaway greenhouse effect, but not so far that low temperatures cause all water on the planet to freeze.

Earth could not have condensed from the protoplanetary disk with its current oceans of water because the early inner Solar System was far too hot for water to condense. Instead, water and other volatiles must have been delivered to Earth from the outer Solar System later in its history. Modern geochemical evidence suggests that water was delivered to Earth by impacts from icy planetesimals similar in composition to modern asteroids in the outer edges of the asteroid belt. However, when and how that water was delivered to Earth is the subject of ongoing research.

Earth's Water Inventory

While the majority of Earth's surface is covered by oceans, those oceans make up just a small fraction of the mass of the planet. The mass of Earth's oceans is estimated to be 1.37×10^{21} kg, which is 0.023% of the total mass of Earth, 6.0×10^{24} kg. An additional 0.5×10^{21} kg of water is estimated to exist in ice, lakes, rivers, groundwater, and atmospheric water vapor. A significant amount of water is also stored in Earth's crust, mantle, and core. Unlike molecular H_2O that is found on the surface, water in the interior exists primarily in hydrated minerals or as trace amounts of hydrogen bonded to oxygen atoms in anhydrous minerals. Hydrated silicates on the surface transport water into the mantle at convergent plate boundaries, where oceanic crust is subducted underneath continental crust. While it is difficult to estimate the total water content of the mantle due to limited samples, approximately three times the mass of the Earth's oceans could be stored there. Similarly, the Earth's core could contain four to five oceans worth of hydrogen.

Hypotheses for the Origins of Earth's Water

Extraplanetary Sources

Water has a much lower condensation temperature than other materials that compose the terrestrial planets in the Solar System, such as iron and silicates. The region of the protoplanetary disk closest to the Sun was very hot early in the history of the Solar System, and it is not feasible that oceans of water condensed with the Earth as it formed. Further from the young Sun where temperatures were cooler, water could condense and form icy planetesimals. The boundary of the region where ice could form in the early Solar System is known as the frost line (or snow line), and is located in the modern asteroid belt, between about 2.7 and 3.1 Astronomical Units (AU) from the Sun. It is therefore necessary that objects forming beyond the frost line–such as comets, trans-Neptunian objects, and water-rich meteoroids (protoplanets)–delivered water to Earth. However, the timing of this delivery is still in question.

One theory claims that Earth accreted (gradually grew by accumulation of) icy planetesimals about 4.5 billion years ago, when it was 60 to 90% of its current size. In this scenario, Earth was able to retain water in some form throughout accretion and major impact events. This hypothesis is supported by similarities in the abundance and the isotope ratios of water between the oldest known carbonaceous chondrite meteorites and meteorites from Vesta, both of which originate from the Solar System's asteroid belt. It is also supported by studies of osmium isotope ratios, which suggest that a sizeable quantity of water was contained in the material that Earth accreted early on. Measurements of the chemical composition of lunar samples collected by the Apollo 15 and 17 missions further support this, and indicate that water was already present on Earth before the Moon was formed.

One problem with this hypothesis is that the noble gas isotope ratios of Earth's atmosphere are different from those of its mantle, which suggests they were formed from different sources. To explain this observation, a so-called "late veneer" theory has been proposed in which water was delivered much later in Earth's history, after the Moon-forming impact. However, our current understanding of Earth's formation allows for less than 1% of Earth's material accreting after the Moon formed, implying that the material accreted later must have been very water-rich. Models of early Solar System dynamics have shown that icy asteroids could have been delivered to the inner Solar System (including Earth) during this period if Jupiter migrated closer to the Sun.

Yet a third hypothesis, supported by evidence from molybdenum isotope ratios, suggests that the Earth gained most of its water from the same interplanetary collision that caused the formation of the Moon.

Geochemical Analysis of Water in the Solar System

Carbonaceous chondrites such as the Allende Meteorite (above) likely delivered much of the Earths water, as evidenced by their isotopic similarities to ocean water.

Isotopic ratios provide a unique "chemical fingerprint" that is used to compare Earth's water with reservoirs elsewhere in the Solar System. One such isotopic ratio, that of deuterium to hydrogen (D/H), is particularly useful in the search for the origin of water on Earth. Hydrogen is the most abundant element in the universe, and its heavier isotope deuterium can sometimes take the place of a hydrogen atom in molecules like H_2O. Most deuterium was created in the Big Bang or in supernovae, so its uneven distribution throughout the protosolar nebula was effectively "locked in" early in the formation of the Solar System. By studying the different isotopic ratios of Earth and of other icy bodies in the Solar System, we can search for the likely origins of Earth's water.

Earth

The deuterium to hydrogen ratio for ocean water on Earth is known very precisely to be $(1.5576 \pm 0.0005) \times 10^{-4}$. This value represents a mixture of all of the sources that contributed to Earth's reservoirs, and is used to identify the source or sources of Earth's water. The ratio of deuterium to hydrogen may have increased over the Earth's lifetime as the lighter isotope is more likely to leak to space in atmospheric loss processes. However no process is known that can decrease Earth's D/H ratio over time. This loss of the lighter isotope is one explanation for why Venus has such a high D/H ratio, as that planet's water was vaporized during the runaway greenhouse effect and subsequently lost much of its hydrogen to space. Because Earth's D/H ratio has increased significantly over time, the D/H ratio of water originally delivered to the planet was lower than what we currently measure. This is consistent with a scenario in which a significant proportion of the water on Earth was already present during the planet's early evolution.

Asteroids

Comet Halley as imaged by the European Space Agency's *Giotto* probe in 1986. *Giotto* flew by Halley's Comet and analyzed the isotopic levels of ice sublimating from the comet's surface using a mass spectrometer.

Multiple geochemical studies have concluded that asteroids are most likely the primary source of Earth's water. Carbonaceous chondrites–which are a subclass of the oldest meteorites in the Solar System–have isotopic levels most similar to ocean water. The CI and CM subclasses of carbonaceous chondrites specifically have hydrogen and nitrogen isotope levels that closely match Earth's seawater, which suggests water in these meteorites could be the source of Earth's oceans. Two 4.5 billion-year-old meteorites found on Earth that contained liquid water alongside a wide diversity of deuterium-poor organic compounds further support this. Earth's current deuterium to hydrogen ratio also matches ancient eucrite chondrites, which originate from the asteroid Vesta in the outer asteroid belt. CI, CM, and eucrite chondrites are believed to have the same water content and isotope ratios as ancient icy protoplanets from the outer asteroid belt that later delivered water to Earth.

Comets

Comets are kilometer-sized bodies made of dust and ice that originate from the Kuiper Belt (20-50 AU) and the Oort Cloud (>5,000 AU), but have highly elliptical orbits which bring them into the inner solar system. Their icy composition and trajectories which bring them into the inner solar system make them a target for remote and *in situ* measurements of D/H ratios.

It is implausible that Earth's water originated only from comets, since isotope measurements of the deuterium to hydrogen (D/H) ratio in comets Halley, Hyakutake, Hale–Bopp, 2002T7, and Tuttle, yield values approximately twice that of oceanic water.

Using this cometary D/H ratio, models predict that less than 10% of Earth's water was supplied from comets.

Other, shorter period comets (<20 years) called Jupiter family comets likely originate from the Kuiper belt, but have had their orbital paths influenced by gravitational interactions with Jupiter or Neptune. 67P/Churyumov–Gerasimenko is one such comet that was the subject of isotopic measurements by the *Rosetta* spacecraft, which found the comet has a D/H ratio three times that of Earth's seawater. Another Jupiter family comet, 103P/Hartley 2, has a D/H ratio which is consistent with Earth's seawater, but its nitrogen isotope levels do not match Earth's.

Hydrothermal Vent

A hydrothermal vent is a fissure on the seafloor from which geothermally heated water issues. Hydrothermal vents are commonly found near volcanically active places, areas where tectonic plates are moving apart at spreading centers, ocean basins, and hotspots. Hydrothermal deposits are rocks and mineral ore deposits formed by the action of hydrothermal vents.

Hydrothermal vents exist because the earth is both geologically active and has large amounts of water on its surface and within its crust. Under the sea, hydrothermal vents may form features called black smokers or white smokers. Relative to the majority of the deep sea, the areas around submarine hydrothermal vents are biologically more productive, often hosting complex communities fueled by the chemicals dissolved in the vent fluids. Chemosynthetic bacteria and archaea form the base of the food chain, supporting diverse organisms, including giant tube worms, clams, limpets and shrimp. Active hydrothermal vents are thought to exist on Jupiter's moon Europa, and Saturn's moon Enceladus, and it is speculated that ancient hydrothermal vents once existed on Mars.

Physical Properties

Hydrothermal vents in the deep ocean typically form along the mid-ocean ridges, such as the East Pacific Rise and the Mid-Atlantic Ridge. These are locations where two tectonic plates are diverging and new crust is being formed.

The water that issues from seafloor hydrothermal vents consists mostly of sea water drawn into the hydrothermal system close to the volcanic edifice through faults and porous sediments or volcanic strata, plus some magmatic water released by the upwelling magma. In terrestrial hydrothermal systems, the majority of water circulated within the fumarole and geyser systems is meteoric water plus ground water that has percolated down into the thermal system from the surface, but it also commonly contains some portion of metamorphic water, magmatic water, and sedimentary formational brine that is released by the magma. The proportion of each varies from location to location.

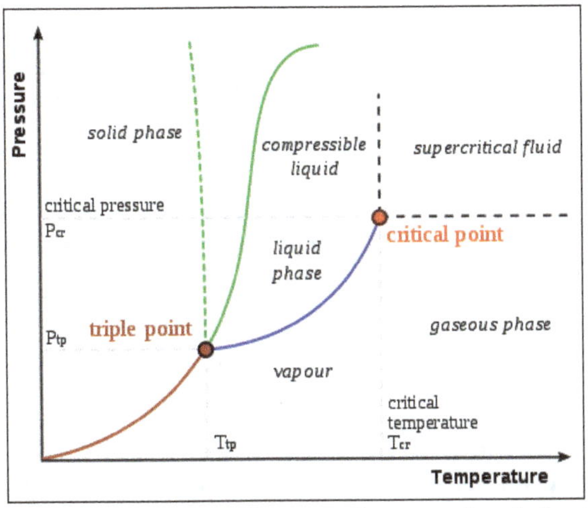

In this phase diagram, the green dotted line illustrates the anomalous behavior of water. The dotted green line marks the melting point and the blue line the boiling point, showing how they vary with pressure; the solid green line shows the typical melting point behavior for other substances.

In contrast to the approximately 2 °C (36 °F) ambient water temperature at these depths, water emerges from these vents at temperatures ranging from 60 °C (140 °F) up to as high as 464 °C (867 °F). Due to the high hydrostatic pressure at these depths, water may exist in either its liquid form or as a supercritical fluid at such temperatures. The critical point of (pure) water is 375 °C (707 °F) at a pressure of 218 atmospheres.

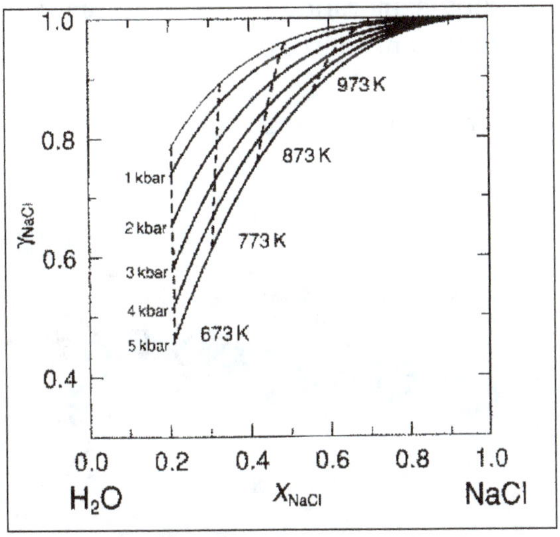

However, introducing salinity into the fluid raises the critical point to higher temperatures and pressures. The critical point of seawater (3.2 wt. % NaCl) is 407 °C (765 °F) and 298.5 bars, corresponding to a depth of ~2,960 m (9,710 ft) below sea level. Accordingly, if a hydrothermal fluid with a salinity of 3.2 wt. % NaCl vents above 407 °C (765 °F) and 298.5 bars, it is supercritical. Furthermore, the salinity of vent fluids have been shown to vary widely due to phase separation in the crust. The critical point for lower salinity fluids is at lower temperature and pressure conditions than that for seawater, but higher than that for pure water. For example, a vent fluid with a 2.24 wt. % NaCl salinity has the critical point at 400 °C (752 °F) and 280.5 bars. Thus, water emerging from the hottest parts of some hydrothermal vents can be a supercritical fluid, possessing physical properties between those of a gas and those of a liquid.

Examples of supercritical venting are found at several sites. *Sister Peak* (Comfortless Cove Hydrothermal Field, 4°48′S 12°22′W4.800°S 12.367°W, depth 2,996 m or 9,829 ft) vents low salinity phase-separated, vapor-type fluids. Sustained venting was not found to be supercritical but a brief injection of 464 °C (867 °F) was well above supercritical conditions. A nearby site, Turtle Pits, was found to vent low salinity fluid at 407 °C (765 °F), which is above the critical point of the fluid at that salinity. A vent site in the Cayman Trough named Beebe, which is the world's deepest known hydrothermal site at ~5,000 m (16,000 ft) below sea level, has shown sustained supercritical venting at 401 °C (754 °F) and 2.3 wt% NaCl.

Although supercritical conditions have been observed at several sites, it is not yet known what significance, if any, supercritical venting has in terms of hydrothermal circulation, mineral deposit formation, geochemical fluxes or biological activity.

The initial stages of a vent chimney begin with the deposition of the mineral anhydrite. Sulfides of copper, iron, and zinc then precipitate in the chimney gaps, making it less porous over the course of time. Vent growths on the order of 30 cm (1 ft) per day have been recorded. An April 2007 exploration of the deep-sea vents off the coast of Fiji found those vents to be a significant source of dissolved iron.

Black Smokers and White Smokers

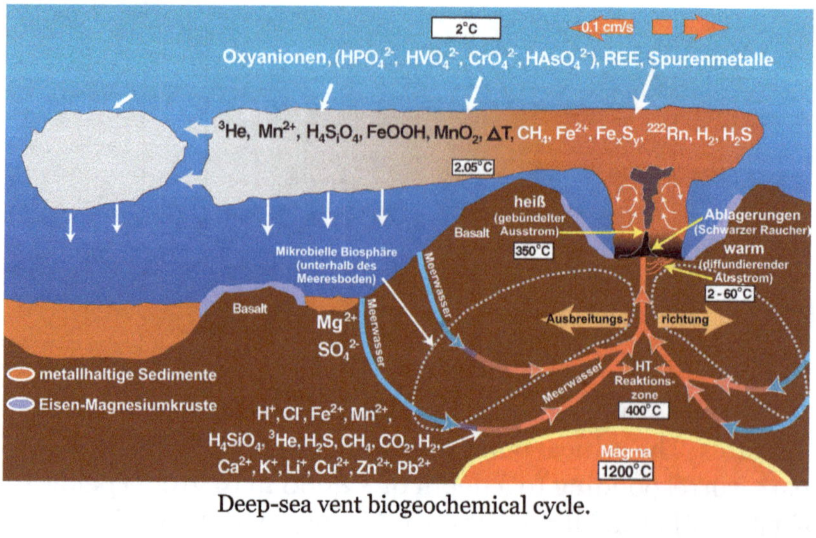

Deep-sea vent biogeochemical cycle.

Some hydrothermal vents form roughly cylindrical chimney structures. These form from minerals that are dissolved in the vent fluid. When the superheated water contacts the near-freezing sea water, the minerals precipitate out to form particles which add to the height of the stacks. Some of these chimney structures can reach heights of 60 m. An example of such a towering vent was "Godzilla", a structure on the Pacific Ocean deep seafloor near Oregon that rose to 40 m before it fell over in 1996.

Black smokers were first discovered in 1979 on the East Pacific Rise at 21° north latitude.

A black smoker or deep sea vent is a type of hydrothermal vent found on the seabed, typically in the bathyal zone (with largest frequency in depths from 2500 m to 3000 m), but also in lesser depths as well as deeper in abyssal zone. They appear as black, chimney-like structures that emit a cloud of black material. Black smokers typically emit particles with high levels of sulfur-bearing minerals, or sulfides. Black smokers are formed in fields hundreds of meters wide when superheated water from below Earth's crust comes through the ocean floor (water may attain temperatures above 400 °C). This water is rich in dissolved minerals from the crust, most notably sulfides. When it comes in contact with cold ocean water, many minerals precipitate, forming a black, chimney-like structure around each vent. The deposited metal sulfides can become massive sulfide ore deposits in time. Some black smokers on the Azores portion of the Mid Atlantic Ridge are extremely rich in metal content, such as Rainbow with 24,000 μM concentrations of iron.

Black smokers were first discovered in 1979 on the East Pacific Rise by scientists from Scripps Institution of Oceanography during the RISE Project. They were observed using the deep submergence vehicle ALVIN from the Woods Hole Oceanographic Institution. Now, black smokers are known to exist in the Atlantic and Pacific Oceans, at an average depth of 2100 metres. The most northerly black smokers are a cluster of five named Loki's Castle, discovered in 2008 by scientists from the University of Bergen at 73°N, on the Mid-Atlantic Ridge between Greenland and Norway. These black smokers are of interest as they are in a more stable area of the Earth's crust, where tectonic forces are less and consequently fields of hydrothermal vents are less common. The world's deepest known black smokers are located in the Cayman Trough, 5,000 m (3.1 miles) below the ocean's surface.

White smoker vents emit lighter-hued minerals, such as those containing barium, calcium and silicon. These vents also tend to have lower-temperature plumes probably because they are generally distant from their heat source.

Black and white smokers may coexist in the same hydrothermal field, but they generally represent proximal and distal vents to the main upflow zone, respectively. However, white smokers correspond mostly to waning stages of such hydrothermal fields, as magmatic heat sources become progressively more distant from the source (due to magma crystallization) and hydrothermal fluids become dominated by seawater instead of magmatic water. Mineralizing fluids from this type of vents are rich in calcium and they form dominantly sulfate-rich (i.e., barite and anhydrite) and carbonate deposits.

Biological Communities

Life has traditionally been seen as driven by energy from the sun, but deep-sea organisms have no access to sunlight, so they must depend on nutrients found in the dusty chemical deposits and hydrothermal fluids in which they live. Previously, benthic oceanographers assumed that vent organisms were dependent on marine snow, as deep-sea organisms are. This would leave them dependent on plant life and thus the sun. Some hydrothermal vent organisms do consume this "rain", but with only such a system, life forms would be sparse. Compared to the surrounding sea floor, however, hydrothermal vent zones have a density of organisms 10,000 to 100,000 times greater.

Hydrothermal vent communities are able to sustain such vast amounts of life because vent organisms depend on chemosynthetic bacteria for food. The water from the hydrothermal vent is rich in dissolved minerals and supports a large population of chemoautotrophic bacteria. These bacteria use sulfur compounds, particularly hydrogen sulfide, a chemical highly toxic to most known organisms, to produce organic material through the process of chemosynthesis.

A dense fauna (*Kiwa* anomurans and *Vulcanolepas*-like stalked barnacles) near East Scotia Ridge vents.

The ecosystem so formed is reliant upon the continued existence of the hydrothermal vent field as the primary source of energy, which differs from most surface life on Earth, which is based on solar energy. However, although it is often said that these communities exist independently of the sun, some of the organisms are actually dependent upon oxygen produced by photosynthetic organisms, while others are anaerobic.

Giant tube worms (*Riftia pachyptila*) cluster around vents in the Galapagos Rift.

The chemosynthetic bacteria grow into a thick mat which attracts other organisms, such as amphipods and copepods, which graze upon the bacteria directly. Larger organisms, such as snails, shrimp, crabs, tube worms, fish (especially eelpout, cutthroat eel, ophidiiforms and *Symphurus thermophilus*), and octopuses (notably *Vulcanoctopus hydrothermalis*), form a food chain of predator and prey relationships above the primary consumers. The main families of organisms found around seafloor vents are annelids, pogonophorans, gastropods, and crustaceans, with large bivalves, vestimentiferan worms, and "eyeless" shrimp making up the bulk of nonmicrobial organisms.

Siboglinid tube worms, which may grow to over 2 m (6.6 ft) tall in the largest species, often form an important part of the community around a hydrothermal vent. They have no mouth or digestive tract, and like parasitic worms, absorb nutrients produced by the bacteria in their tissues. About 285 billion bacteria are found per ounce of tubeworm tissue. Tubeworms have red plumes which contain hemoglobin. Hemoglobin combines with hydrogen sulfide and transfers it to the bacteria living inside the worm. In return, the bacteria nourish the worm with carbon compounds. Two of the species that inhabit a hydrothermal vent are *Tevnia jerichonana*, and *Riftia pachyptila*. One discovered community, dubbed "Eel City", consists predominantly of the eel *Dysommina rugosa*. Though eels are not uncommon, invertebrates typically dominate hydrothermal vents. Eel City is located near Nafanua volcanic cone, American Samoa.

In 1993, already more than 100 gastropod species were known to occur in hydrothermal vents. Over 300 new species have been discovered at hydrothermal vents, many of them "sister species" to others found in geographically separated vent areas. It has been proposed that before the North American plate overrode the mid-ocean ridge, there was a single biogeographic vent region found in the eastern Pacific. The subsequent barrier to travel began the evolutionary divergence of species in different locations. The examples of convergent evolution seen between distinct hydrothermal vents is seen as major support for the theory of natural selection and of evolution as a whole.

Although life is very sparse at these depths, black smokers are the centers of entire ecosystems. Sunlight is nonexistent, so many organisms – such as archaea and extremophiles – convert the heat, methane, and sulfur compounds provided by black smokers into energy through a process called chemosynthesis. More complex life forms, such as clams and tubeworms, feed on these organisms. The organisms at the base of the food chain also deposit minerals into the base of the black smoker, therefore completing the life cycle.

A species of phototrophic bacterium has been found living near a black smoker off the coast of Mexico at a depth of 2,500 m (8,200 ft). No sunlight penetrates that far into the waters. Instead, the bacteria, part of the Chlorobiaceae family, use the faint glow from the black smoker for photosynthesis. This is the first organism discovered in nature to exclusively use a light other than sunlight for photosynthesis.

New and unusual species are constantly being discovered in the neighborhood of black smokers. The Pompeii worm *Alvinella pompejana*, which is capable of withstanding temperatures up to 80 °C (176 °F), was found in the 1980s, and a scaly-foot gastropod *Chrysomallon squamiferum* in 2001 during an expedition to the Indian Ocean's Kairei hydrothermal vent field. The latter uses iron sulfides (pyrite and greigite) for the structure of its dermal sclerites (hardened body parts), instead of calcium carbonate. The extreme pressure of 2500 m of water (approximately 25 megapascals or 250 atmospheres) is thought to play a role in stabilizing iron sulfide for biological purposes. This armor plating probably serves as a defense against the venomous radula (teeth) of predatory snails in that community.

In March 2017, researchers reported evidence of possibly the oldest forms of life on Earth. Putative fossilized microorganisms were discovered in hydrothermal vent precipitates in the Nuvvuagittuq Belt of Quebec, Canada, that may have lived as early as 4.280 billion years ago, not long after the oceans formed 4.4 billion years ago, and not long after the formation of the Earth 4.54 billion years ago.

Animal-Bacterial Symbiosis

Hydrothermal vent ecosystems have enormous biomass and productivity; but this rests on the symbiotic relationships that have evolved at vents. Deep-sea hydrothermal vent ecosystems differ from their shallow-water and terrestrial hydrothermal counterparts due to the symbiosis that occurs between macro invertebrate hosts and chemoautotrophic microbial symbionts in the former. Since sunlight does not reach deep-sea hydrothermal vents, organisms in deep-sea hydrothermal vents cannot obtain energy from the sun to perform photosynthesis. Instead, the microbial life found at hydrothermal vents are chemosynthetic; they fix carbon by using energy from chemicals such as sulfide, as opposed to light energy from the sun. In other words, the symbiont converts inorganic molecules (H_2S, CO_2, O) to organic molecules that the host then uses as nutrition. However, sulfide is an extremely toxic substance to most life on Earth. For this reason, scientists were astounded when they first found hydrothermal vents teeming with life in 1977. What was discovered was the ubiquitous symbiosis of chemoautotrophs living in (endosymbiosis) the vent animals' gills; the reason why multicellular life is capable to survive the toxicity of vent systems. Scientists are therefore now studying how the microbial symbionts aid in sulfide detoxification (therefore allowing the host to survive the otherwise toxic conditions). Work on microbiome function shows that host-associated microbiomes are also important in host development, nutrition, defense against predators, and detoxification. In return, the host provides the symbiont with chemicals required for chemosynthesis, such as carbon, sulfide, and oxygen.

In the early stages of studying life at hydrothermal vents, there were differing theories regarding the mechanisms by which multicellular organisms were able to acquire nutrients from these environments, and how they were able to survive in such extreme conditions. In 1977, it was hypothesized that the chemoautotrophic bacteria at hydrothermal vents might be responsible for contributing to the diet of suspension feeding bivalves.

Finally, in 1981, it was understood that giant tubeworm nutrition acquisition occurred as a result of chemoautotrophic bacterial endosymbionts. As scientists continued to study life at hydrothermal vents, it was understood that symbiotic relationships between chemoautotrophs and macrofauna invertebrate species was ubiquitous. For instance, in 1983, clam gill tissue was confirmed to contain bacterial endosymbionts; in 1984 vent bathymodiolid mussels and vesicomyid clams were also found to carry endosymbionts.

However, the mechanisms by which organisms acquire their symbionts differ, as do the metabolic relationships. For instance, tubeworms have no mouth and no gut, but they do have a "trophosome", which is where they deal with nutrition and where their endosymbionts are found. They also have a bright red plume, which they use to uptake compounds such as O, H_2S, and CO_2, which feed the endosymbionts in their trophosome. Remarkably, the tubeworms hemoglobin (which incidentally is the reason for the bright red color of the plume) is capable of carrying oxygen without interference or inhibition from sulfide, despite the fact that oxygen and sulfide are typically very reactive. In 2005, it was discovered that this is possible due to zinc ions that bind the hydrogen sulfide in the tubeworms hemoglobin, therefore preventing the sulfide from reacting with the oxygen. It also reduces the tubeworms tissue from exposure to the sulfide and provides the bacteria with the sulfide to perform chemoautotrophy. It has also been discovered that tubeworms can metabolize CO_2 in two different ways, and can alternate between the two as needed as environmental conditions change.

In 1988, research confirmed thiotrophic (sulfide-oxidizing) bacteria in Alvinochonca hessleri, a large vent mollusk. In order to circumvent the toxicity of sulfide, mussels first convert it to thiosulfate before carrying it over to the symbionts. In the case of motile organisms such as alvinocarid shrimp, they must track oxic (oxygen-rich) / anoxic (oxygen-poor) environments as they fluctuate in the environment.

Organisms living at the edge of hydrothermal vent fields, such as pectinid scallops, also carry endosymbionts in their gills, and as a result their bacterial density is low relative to organisms living nearer to the vent. However, the scallop's dependence on the microbial endosymbiont for obtaining their nutrition is therefore also lessened.

Furthermore, not all host animals have endosymbionts; some have episymbionts—symbionts living on the animal as opposed to inside the animal. Shrimp found at vents in the Mid-Atlantic Ridge were once thought of as an exception to the necessity of symbiosis for macroinvertebrate survival at vents. That changed in 1988 when they were discovered to carry episymbionts. Since then, other organisms at vents have been found to carry episymbionts as well, such as Lepetodrilis fucensis.

Furthermore, while some symbionts reduce sulfur compounds, others are known as "methanotrophs" and reduce carbon compounds, namely methane. Bathmodiolid mussels are an example of a host that contains methanotrophic endosymbionts; however, the latter mostly occur in cold seeps as opposed to hydrothermal vents.

While chemosynthesis occurring at the deep ocean allows organisms to live without sunlight in the immediate sense, they technically still rely on the sun for survival, since oxygen in the ocean is a byproduct of photosynthesis. However, if the sun were to suddenly disappear and photosynthesis ceased to occur on our planet, life at the deep-sea hydrothermal vents could continue for millennia (until the oxygen was depleted).

Biological Theories

The Deep Hot Biosphere

Thomas Gold referred to *ocean vents* in support of his theory that the lower levels of the earth are rich in living biological material that finds its way to the surface. He further expanded his ideas in the book "The Deep Hot Biosphere".

Hydrothermal Origin of Life

The chemical and thermal dynamics in hydrothermal vents makes such environments highly suitable thermodynamically for chemical evolution processes to take place. Therefore, thermal energy flux is a permanent agent and contributed to the evolution of the planet, including prebiotic chemistry.

Günter Wächtershäuser proposed the iron-sulfur world theory and suggested that life might have originated at hydrothermal vents. Wächtershäuser proposed that an early form of metabolism predated genetics. By metabolism he meant a cycle of chemical reactions that release energy in a form that can be harnessed by other processes.

It has been proposed that amino acid synthesis could have occurred deep in the Earth's crust and that these amino acids were subsequently shot up along with hydrothermal fluids into cooler waters, where lower temperatures and the presence of clay minerals would have fostered the formation of peptides and protocells. This is an attractive hypothesis because of the abundance of CH_4 (methane) and NH_3 (ammonia) present in hydrothermal vent regions, a condition that was not provided by the Earth's primitive atmosphere. A major limitation to this hypothesis is the lack of stability of organic molecules at high temperatures, but some have suggested that life would have originated outside of the zones of highest temperature. There are numerous species of extremophiles and other organisms currently living immediately around deep-sea vents, suggesting that this is indeed a possible scenario.

Experimental research and computing modeling indicate that the surfaces of mineral particles inside hydrothermal vents have similar catalytic properties to enzymes and are able to create simple organic molecules, such as methanol (CH_3OH) and formic acid (HCO_2H), out of the dissolved CO_2 in the water.

It is thought that alkaline hydrothermal vents (white smokers) might be more suitable for emerging life than black smokers due to their pH conditions.

Distribution

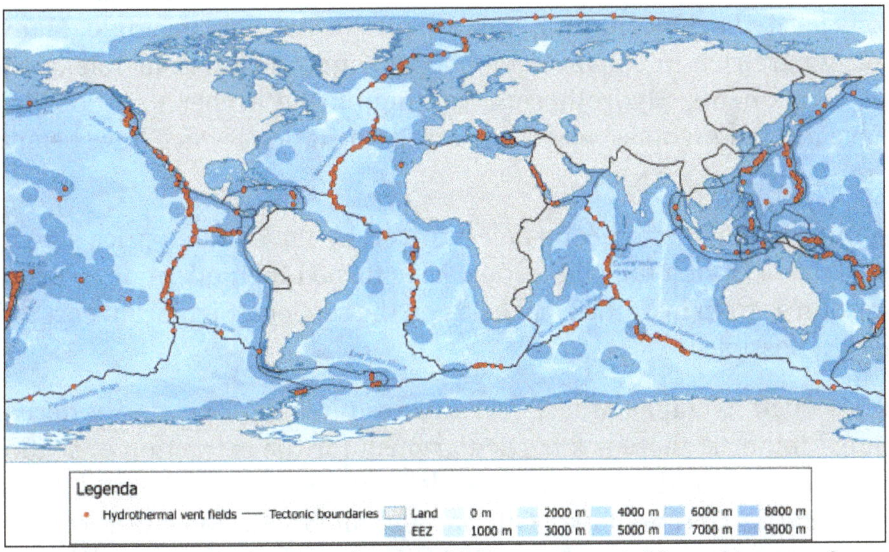

Distribution of hydrothermal vents. This map was created by making use of the InterRidge ver.3.3 database.

Hydrothermal vents tend to be distributed along the Earth's plate boundaries, although they may also be found at intra-plate locations such as hotspot volcanoes. As of 2009 there were approximately 500 known active submarine hydrothermal vent fields, with about half visually observed at the seafloor and the other half suspected from water column indicators and seafloor deposits. The InterRidge program office hosts a global database for the locations of known active submarine hydrothermal vent fields.

Rogers recognized at least 11 biogeographic provinces of hydrothermal vent systems:

1. Mid-Atlantic Ridge province,
2. East Scotia Ridge province,
3. Northern East Pacific Rise province,
4. Central East Pacific Rise province,
5. Southern East Pacific Rise province,
6. South of the Easter Microplate,
7. Indian Ocean province,
8. Four provinces in the western Pacific.

Exploitation

Hydrothermal vents, in some instances, have led to the formation of exploitable mineral resources via deposition of seafloor massive sulfide deposits. The Mount Isa orebody located in Queensland, Australia, is an excellent example. Many hydrothermal vents are rich in cobalt, gold, copper, and rare earth metals essential for electronic components. Hydrothermal venting on the Archean seafloor is considered to have formed Algoma-type banded iron formations which have been a source of iron ore.

Recently, mineral exploration companies, driven by the elevated price activity in the base metals sector during the mid-2000s, have turned their attention to extraction of mineral resources from hydrothermal fields on the seafloor. Significant cost reductions are, in theory, possible.

In countries such as Japan where mineral resources are primarily derived from international imports, there is a particular push for the extraction of seafloor mineral resources. The world's first "large-scale" mining of hydrothermal vent mineral deposits was carried out by Japan Oil, Gas and Metals National Corporation (JOGMEC) in August - September, 2017. JOGMEC carried out this operation using the Research Vessel Hakurei. This mining was carried out at the 'Izena hole/cauldron' vent field within the hydrothermally active back-arc basin known as the Okinawa Trough which contains 15 confirmed vent fields according to the InterRidge Vents Database.

Two companies are currently engaged in the late stages of commencing to mine seafloor massive sulfides. Nautilus Minerals is in the advanced stages of commencing extraction from its Solwarra deposit, in the Bismarck Archipelago, and Neptune Minerals is at an earlier stage with its Rumble II West deposit, located on the Kermadec Arc, near the Kermadec Islands. Both companies are proposing using modified existing technology. Nautilus Minerals, in partnership with Placer Dome (now part of Barrick Gold), succeeded in 2006 in returning over 10 metric tons of mined SMS to the surface using modified drum cutters mounted on an ROV, a world first. Neptune Minerals in 2007 succeeded in recovering SMS sediment samples using a modified oil industry suction pump mounted on an ROV, also a world first.

Potential seafloor mining has environmental impacts including dust plumes from mining machinery affecting filter-feeding organisms, collapsing or reopening vents, methane clathrate release, or even sub-oceanic land slides. A large amount of work is currently being engaged in by both the above-mentioned companies to ensure that potential environmental impacts of seafloor mining are well understood and control measures are implemented, before exploitation commences. However, this process has been arguably hindered by the disproportionate distribution of research effort among vent ecosystems: the best studied and understood hydrothermal vent ecosystems are not representative of those targeted for mining.

Attempts have been made in the past to exploit minerals from the seafloor. The 1960s and 70s saw a great deal of activity (and expenditure) in the recovery of manganese nodules from the abyssal plains, with varying degrees of success. This does demonstrate however that recovery of minerals from the seafloor is possible, and has been possible for some time. Mining of manganese nodules served as a cover story for the elaborate attempt in 1974 by the CIA to raise the sunken Soviet submarine *K-129*, using the Glomar Explorer, a ship purpose built for the task by Howard Hughes. The operation was known as Project Azorian, and the cover story of seafloor mining of manganese nodules may have served as the impetus to propel other companies to make the attempt.

Conservation

The conservation of hydrothermal vents has been the subject of sometimes heated discussion in the oceanographic community for the last 20 years. It has been pointed out that it may be that those causing the most damage to these fairly rare habitats are scientists. There have been attempts to forge agreements over the behaviour of scientists investigating vent sites but although there is an agreed code of practice there is as yet no formal international and legally binding agreement.

SIGNIFICANCE OF WATER IN BIOSPHERE

The cyclic movement of water, permeating different spheres of the ecosystem, is as vital and important as energy. Different states of water—solid, liquid and gas—allow its easy passage through hydro- litho-atmosphere.

The cyclic movement of water through the hydrosphere, atmosphere, lithosphere and biosphere is represented by the hydrological cycle. The processes involved in the hydrological cycle are evaporation, condensation, precipitation, run-off, seepage and transpiration. While evaporation, condensation and precipitation are atmospheric processes, run-off and seepage are geomorphic processes. Transpiration is the process of the biosphere.

The hydrological cycle is of great importance, as it is essential for all forms of life. It fulfills the following functions:

- Water has conditioned the ecology and evolution of all terrestrial organisms. These organisms either adapt themselves to seasonal deficiencies of water or store water in their body parts.
- Water is a major constituent of all organisms. Water makes up 99% of some marine invertebrates, 85% of some mammals and 66% of most vertebrates.
- Water supplies hydrogen, for the formation of organic compounds in the process of photosynthesis, and oxygen to the air.
- Water acts as a solvent for mineral nutrients which are absorbed by plants through their roots.
- Water is a circulatory medium for carrying nutrients and for carrying away waste products of metabolism.
- It is not just the vegetation, but the biotic composition of the soil, soil chemistry and composition of various minerals supplied to the organic matter, which are controlled by water present in them.

Addition and Loss of Moisture

Moisture is added to the soil by the processes of precipitation and infiltration. And moisture is lost by evaporation and run-off in the form of streams and rivers. Loss of moisture also occurs by gravity percolation to the ground water zone below the soil layer.

Soil moisture is also lost by transpiration from the leaves of plants. In any land plant, as much as 98% of the water taken by roots passes through and gets lost by transpiration. Thus, it is evaporation which is responsible for total loss of water from the surface.

The water available for living organisms on the lithosphere is a small fraction of the total water that falls on the surface of the earth. The water needs of plants vary, depending on the type of plant and its size. The soil moisture budget in any area shows seasonal variations depending upon temperature and precipitation in each season.

There are seasons of soil moisture surplus and deficiency and the plants have to make physiological adaptations to overcome seasons of moisture deficiency.

BIOGENIC SUBSTANCE

A biogenic substance is a product made by or of life forms. The term encompasses constituents, secretions, and metabolites of plants or animals. Biogenic substances are referred to as biomolecules.

Examples of Biogenic Substance:

- Coal and oil are possible examples of constituents which may have undergone changes over geologic time periods.
- Chalk and limestone are examples of secretions (marine animal shells) which are of geologic age.
- Grass and wood are biogenic constituents of contemporary origin.
- Pearls, silk and ambergris are examples of secretions of contemporary origin.
- Biogenic neurotransmitters.

Abiogenic

An abiogenic substance or process does not result from the present or past activity of living organisms. Abiogenic products may, e.g., be minerals, other inorganic compounds, as well as simple organic compounds (e.g. extraterrestrial methane).

Mineral

A mineral is, broadly speaking, a solid chemical compound that occurs in pure form. A rock may consist of a single mineral, or may be an aggregate of two or more different minerals, spacially segregated into distinct phases. Compounds that occur only in living beings are excluded, but some minerals are biogenic and/or are organic compounds in the sense of chemistry. Moreover, living beings synthesize inorganic minerals that occur in rocks. In geology and mineralogy, the term "mineral" is reserved for mineral species: crystalline compounds with a well-defined chemical composition and a specific crystal structure. Minerals without a definite crystalline structure, such as opal or obsidian, are more properly called mineraloids. If a chemical compound may occur with different crystal structures, each structure is considered different mineral species. Thus, for example and stishovite are two different minerals consisting of the same compound, silicon dioxide; the International Mineralogical

Association is the world's premier standard body for the definition and nomenclature of mineral species.

As of November 2018, the IMA recognizes 5,413 official mineral species. Out of more than 5,500 proposed or traditional ones; the chemical composition of a named mineral species may vary somewhat by the inclusion of small amounts of impurities. Specific varieties of a species sometimes have official names of their own. For example, amethyst is a purple variety of the mineral species quartz; some mineral species can have variable proportions of two or more chemical elements that occupy equivalent positions in the mineral's structure. Sometimes a mineral with variable composition is split into separate species, more or less arbitrarily, forming a mineral group. Besides the essential chemical composition and crystal structure, the description of a mineral species includes its common physical properties such as habit, lustre, colour, tenacity, fracture, specific gravity, fluorescence, radioactivity, as well as its taste or smell and its reaction to acid. Minerals are classified by key chemical constituents.

Silicate minerals comprise 90% of the Earth's crust. Other important mineral groups include the native elements, oxides, carbonates and phosphates. One definition of a mineral encompasses the following criteria: Formed by a natural process. Stable or metastable at room temperature. Classical examples of exceptions to this rule include native mercury, which crystallizes at −39 °C, water ice, solid only below 0 °C. Modern advances have included extensive study of liquid crystals, which extensively involve mineralogy. Represented by a chemical formula. Minerals are chemical compounds, as such they can be described by fixed or a variable formula. Many mineral groups and species are composed of a solid solution. For example, the olivine group is described by the variable formula 2SiO4, a solid solution of two end-member species, magnesium-rich forsterite and iron-rich fayalite, which are described by a fixed chemical formula. Mineral species themselves could have a variable composition, such as the sulfide mackinawite, 9S8, a ferrous sulfide, but has a significant nickel impurity, reflected in its formula.

Ordered atomic arrangement. This means crystalline. An ordered atomic arrangement gives rise to a variety of macroscopic physical properties, such as crystal form and cleavage. There have been several recent proposals to classify amorphous substances as minerals; the formal definition of a mineral approved by the IMA in 1995: "A mineral is an element or chemical compound, crystalline and, formed as a result of geological processes." Biogenic substances are explicitly excluded by the IMA: "Biogenic substances are chemical compounds produced by biological processes without a geological component and are not regarded as minerals. However, if geological processes were involved in the genesis of the compound the product can be accepted as a mineral". The first three general characteristics are less debated than the last two. Mineral classification schemes and their definitions are evolving to match recent advances in mineral science. Recent changes have included the addition of an organic class, in both the new Dana and the Strunz classification schemes.

The organic class includes a rare group of minerals with hydrocarbons. The IMA Commission on New Minerals and Mineral Names adopted in 2009 a hierarchical scheme for the naming and classification of mineral groups and group names and established seven commissions and four working groups to review and classify minerals into an official listing of their published names.

Silk

Silk is a natural protein fiber, some forms of which can be woven into textiles. The protein fiber of silk is composed of fibroin and is produced by certain insect larvae to form cocoons; the best-known silk is obtained from the cocoons of the larvae of the mulberry silkworm Bombyx mori reared in captivity. The shimmering appearance of silk is due to the triangular prism-like structure of the silk fibre, which allows silk cloth to refract incoming light at different angles, thus producing different colors. Silk is produced by several insects. There has been some research into other types of silk. Silk is produced by the larvae of insects undergoing complete metamorphosis, but some insects, such as webspinners and raspy crickets, produce silk throughout their lives. Silk production occurs in Hymenoptera, mayflies, leafhoppers, lacewings, fleas and midges. Other types of arthropods produce most notably various arachnids, such as spiders; the word silk comes from Old English.

Sērikós, "silken" from an Asian source — compare Mandarin sī "silk", Manchurian sirghe, Mongolian sirkek. Several kinds of wild silk, which are produced by caterpillars other than the mulberry silkworm, have been known and used in China, South Asia, Europe since ancient times. However, the scale of production was always far smaller than for cultivated silks. There are several reasons for this: first, they differ from the domesticated varieties in colour and texture and are therefore less uniform. Thus, the only way to obtain silk suitable for spinning into textiles in areas where commercial silks are not cultivated was by tedious and labor-intensive carding. Commercial silks originate from reared silkworm pupae, which are bred to produce a white-colored silk thread with no mineral on the surface; the pupae are killed by either dipping them in boiling water before the adult moths emerge or by piercing them with a needle. These factors all contribute to the ability of the whole cocoon to be unravelled as one continuous thread, permitting a much stronger cloth to be woven from the silk.

Wild silks tend to be more difficult to dye than silk from the cultivated silkworm. A technique known as demineralizing allows the mineral layer around the cocoon of wild silk moths to be removed, leaving only variability in color as a barrier to creating a commercial silk industry based on wild silks in the parts of the world where wild silk moths thrive, such as in Africa and South America. Silk was first developed in ancient China; the earliest example of silk has been found in tombs at the neolithic site Jiahu in Henan, dates back 8,500 years. Silk fabric from 3630 BC was used as wrapping for the body of a child from a Yangshao culture site in Qingtaicun at Xingyang, Henan. Legend gives credit for developing silk to Leizu. Silks were reserved for the Emperors of China for their own use and gifts to others, but spread through Chinese culture and trade both geographically and and to many regions of Asia; because of its texture and lustre, silk became a popular luxury fabric in the many areas accessible to Chinese merchants.

Silk was in great demand, became a staple of pre-industrial international trade. In July 2007, archaeologists discovered intricately woven and dyed silk textiles in a tomb in Jiangxi province, dated to the Eastern Zhou Dynasty 2,500 years ago. Although historians have suspected a long history of a formative textile industry in ancient China, this find of silk textiles employing "complicated techniques" of weaving and dyeing provides direct evidence for silks dating before the Mawangdui-discovery and other silks dating to the Han Dynasty. Silk is described in a chapter of the Fan Shengzhi shu from the Western Han. There is a surviving calendar for silk production in an Eastern

Han document; the two other known works on silk from the Han period are lost. The first evidence of the long distance silk trade is the finding of silk in the hair of an Egyptian mummy of the 21st dynasty, c.1070 BC. The silk trade reached as far as the Indian subcontinent, the Middle East and North Africa; this trade was so extensive that the major set of trade routes between Europe and Asia came to be known as the Silk Road.

The Emperors of China strove to keep knowledge of sericulture secret to maintain the Chinese monopoly. Nonetheless sericulture reached Korea with technological aid from China around 200 BC, the ancient Kingdom of Khotan by AD 50, India by AD 140. In the ancient era, silk from China was the most lucrative and sought-after luxury item traded across the Eurasian continent, many civilizations, such as the ancient Persians, benefited economically from trade. Chinese silk making process: Silk has a long history in India. It is known as Resham in eastern and north India. Pattu in southern parts of India. Recent archaeological discoveries in Harappa and Chanhu-daro suggest that sericulture, employing wild silk threads from native silkworm species, existed in South Asia during the time of the Indus Valley Civilization dating between 2450 BC and 2000 BC, while "hard and fast evidence" for silk production in China dates back to around 2570 BC.

Chemical Substance

A chemical substance is a form of matter having constant chemical composition and characteristic properties. It cannot be separated into components by physical separation methods, i.e. without breaking chemical bonds. Chemical substances can be chemical compounds, or alloys. Chemical elements may not be included in the definition, depending on expert viewpoint. Chemical substances are called 'pure' to set them apart from mixtures. A common example of a chemical substance is pure water. Other chemical substances encountered in pure form are diamond, table salt and refined sugar. However, in practice, no substance is pure, chemical purity is specified according to the intended use of the chemical. Chemical substances exist as solids, gases, or plasma, may change between these phases of matter with changes in temperature or pressure. Chemical substances may be converted to others by means of chemical reactions. Forms of energy, such as light and heat, are not matter, are thus not "substances" in this regard.

A chemical substance may well be defined as "any material with a definite chemical composition" in an introductory general chemistry textbook. According to this definition a chemical substance can either be a pure chemical element or a pure chemical compound. But, there are exceptions to this definition; the chemical substance index published by CAS includes several alloys of uncertain composition. Non-stoichiometric compounds are a special case that violates the law of constant composition, for them, it is sometimes difficult to draw the line between a mixture and a compound, as in the case of palladium hydride. Broader definitions of chemicals or chemical substances can be found, for example: "the term 'chemical substance' means any organic or inorganic substance of a particular molecular identity, including – any combination of such substances occurring in whole or in part as a result of a chemical reaction or occurring in nature". In geology, substances of uniform composition are called minerals, while physical mixtures of several minerals are defined as rocks.

Many minerals, mutually dissolve into solid solutions, such that a single rock is a uniform substance despite being a mixture in stoichiometric terms. Feldspars are a common example: anorthoclase is an alkali aluminum silicate, where the alkali metal is interchangeably either sodium or potassium. In law, "chemical substances" may include both pure substances and mixtures with a defined composition or manufacturing process. For example, the EU regulation REACH defines "monoconstituent substances", "multiconstituent substances" and "substances of unknown or variable composition"; the latter two consist of multiple chemical substances. For example, charcoal is an complex polymeric mixture that can be defined by its manufacturing process. Therefore, although the exact chemical identity is unknown, identification can be made to a sufficient accuracy; the CAS index includes mixtures. Polymers always appear as mixtures of molecules of multiple molar masses, each of which could be considered a separate chemical substance.

However, the polymer may be defined by a known precursor or reaction and the molar mass distribution. For example, polyethylene is a mixture of long chains of -CH_2- repeating units, is sold in several molar mass distributions, LDPE, MDPE, HDPE and UHMWPE; the concept of a "chemical substance" became established in the late eighteenth century after work by the chemist Joseph Proust on the composition of some pure chemical compounds such as basic copper carbonate. He deduced; this is now known as the law of constant composition. With the advancement of methods for chemical synthesis in the realm of organic chemistry. However, there are some controversies regarding this definition because the large number of chemical substances reported in chemistry literature need to be indexed. Isomerism caused much consternation to early researchers, since isomers have the same composition, but differ in configuration of the atoms. For example, there was much speculation for the chemical identity of benzene, until the correct structure was described by Friedrich August Kekulé.

The idea of stereoisomerism that atoms have rigid three-dimensional structure and can thus form isomers that differ only in their three-dimensional arrangement – was another crucial step in understanding the concept of distinct chemical substances. For example, tartaric acid has three distinct isomers, a pair of diastereomers with one diastereomer forming two enantiomers. An element is a chemical substance made up of a particular kind of atom and hence cannot be broken down or transformed by a chemical reaction into a different element, though it can be transmuted into another element through a nuclear reaction.

Abiogenesis

Abiogenesis, or informally the origin of life, is the natural process by which life has arisen from non-living matter, such as simple organic compounds.

MICROBIAL MAT

Microbial mats are benthic small-scale ecosystems that generally develop under environmental conditions that exclude fauna, often referred to as "extreme environments." The biogeochemical cycles in microbial mats are usually largely closed, although small fluxes of elements are exchanged with the geo-, bio-, and atmosphere. An important feature of microbial mats is their carbon and nitrogen autotrophy, i.e., the fixation of inorganic carbon and atmospheric nitrogen are key processes. In illuminated environments, therefore, the photoautotrophic cyanobacteria are the mat-building organisms while in the deep sea or in some caves, chemosynthetic bacteria and archaea are the primary producers.

The term microbial mat originates from the macroscopic structure that these small-scale microbial ecosystems often adopt. Large filamentous microorganisms grow and form an entangled mass with the sediment particles on which they grow, stabilizing the sediment surface and themselves from eroding forces. Mature microbial mats can sometimes be lifted in large pieces from the sediment, just like a doormat. However, in the course of the research on microbial mats, the term has been used for any benthic microbial community that forms a macroscopic structured entity. An important driver for the study of microbial mats is that these systems are considered to be analogues of the Precambrian stromatolites, which have been recognized as the oldest known ecosystems on Earth, and their study could lift the veil of the evolution of early life.

Microbial mats often develop in environments and habitats that are characterized by extreme conditions that exclude metazoans that would otherwise graze on the microbes, preventing them from forming a microbial mat. Microbial mats are largely closed ecosystems. This means that the biogeochemical cycles are basically closed. Therefore, a basic characteristic of microbial mats is that they are carbon autotrophic, meaning that they grow by fixing inorganic carbon.

(a) Microbial mats growing on a sandy intertidal beach. (b) Cross-section of this mat shows the typical varicolored layers with a purple layer of anoxygenic phototrophic bacteria right underneath the layer of cyanobacteria.

This can be done in two principally different ways, namely, by photosynthesis and by chemosynthesis. In the former, light is the source of energy and phototrophic organisms are the organisms that build the microbial mats. In chemosynthetic microbial mats, the source of energy is the oxidation of a reduced compound, often sulfide or methane. In most illuminated environments, the phototrophic cyanobacteria are the mat-building organisms. In dark environments such as in the abyssal of the oceans or in caves, hyperthermal springs, mud volcanoes, and cold seeps provide reduced compounds produced by the Earth's internal heat that drive chemosynthesis. Chemosynthetic mats may also develop in illuminated environments when, for instance, the decomposition of large amounts of organic matter in seawater produces high amounts of sulfide which is in direct contact with oxygen.

Since Precambrian stromatolites are known to have developed in shallow, illuminated coastal environments, the study of coastal microbial mats deserves special attention. Microbial mats develop in coastal intertidal sandy sediments. These environments can be considered as extreme because they are periodically inundated, desiccated, hyper- or hyposaline, experiencing high and low temperatures, and as a rule nutrient poor. Cyanobacteria thrive well under such conditions. As photoautotrophs, they use light, water, and CO_2 as source of energy, electrons, and carbon, respectively, which are abundantly available. Many cyanobacteria are also nitrogen autotrophs, i.e., they can utilize atmospheric N_2, the most abundant form of nitrogen on Earth and after carbon the second most important element for life. Cyanobacteria have low nutrient requirements, possess high-affinity phosphate uptake systems, and contain a variety of storage compounds.

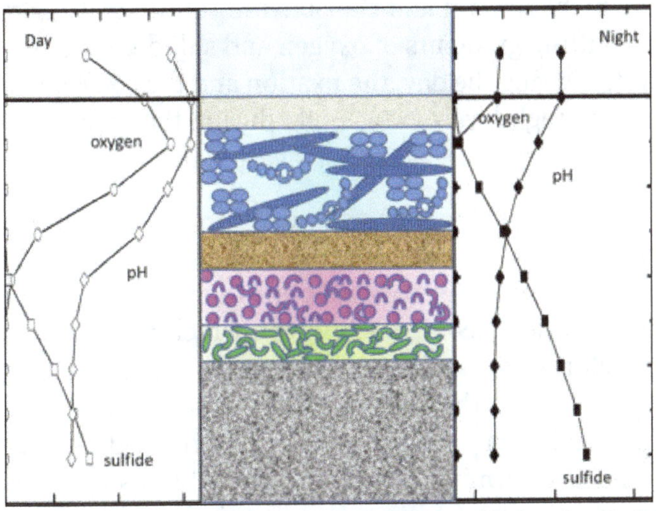

Cartoon of a microbial mat indicating the vertically stratified and varicolored layers of functional groups of microorganisms and the chemical gradients they produce during the day (left) and during the night (right).

The carbon fixed by the cyanobacteria is made available to the microbial community by a variety of different ways. In the light, photorespiration may result in the excretion of glycolate while in the dark, the storage compound glycogen may be fermented to low-molecular compounds such as acetate, lactate, and ethanol. Cyanobacteria exude also exopolymeric substances and upon salinity down shock may release compatible solutes such as the carbohydrates trehalose and sucrose and the polyol glucosylglycerol. Finally, cyanobacteria may die and lyse, providing another source of organic matter and nutrients. A plethora of microorganisms will degrade these compounds, but the sulfate-reducing bacteria deserve a particular notice in marine microbial mats. At night, when oxygenic photosynthesis ceases or in the deeper aphotic layers of the mat, oxygen is quickly consumed by the microbial community, including the cyanobacteria. This happens often within minutes. Sulfate-reducing bacteria are therefore responsible for the bulk of the degradation of organic matter using the abundant sulfate (28 mm) in seawater as electron acceptor, reducing it to sulfide. The sulfide is subsequently oxidized in two ways. In the light, anoxygenic phototrophic purple sulfur bacteria use sulfide as the electron donor for photosynthesis. They oxidize it to elemental sulfur and eventually back to sulfate. These bacteria are basically anaerobic and form a pink layer below the cyanobacteria where they experience anaerobic conditions while sufficient light (particularly of the far red wavelengths) reaches them. The other group of microorganisms that oxidizes the sulfide is the colorless sulfur bacteria. They use oxygen or nitrate as electron acceptor. As a result of the microbial activities, microbial mats are characterized by steep and fluctuating physicochemical gradients. Only during the day, oxygenic photosynthesis by the cyanobacteria enriches and supersaturates the sediment with oxygen. During the night, this oxygen is quickly respired. Sulfide accumulates during the night because light and oxygen (and usually nitrate as well) are lacking for its oxidation while the sulfate-reducing bacteria are continuing to produce it. During the day, sulfide is oxidized. The day-night rhythm and the activities of the mat microorganisms result in the continuously shifting gradients of oxygen and sulfide. Together with that, the pH shows dramatic shifts. During the day, the fixation and the consequent depletion of CO_2 causes the pH to rise to high levels (>9), while during the night, the decomposition of organic matter causes the level of CO_2 to increase and the pH to decrease.

The mat-forming organisms take their positions in these physicochemical gradients. The aerobic cyanobacteria form a green layer at the surface while the anaerobic purple sulfur bacteria form a pink layer right below them. Often, a rusty layer separates them. This layer is supposedly composed of iron which forms a barrier for both oxygen and sulfide, which are toxic to the purple sulfur bacteria and cyanobacteria, respectively. Sulfide diffusing to the surface reacts with iron and oxidizes to elemental sulfur and forms insoluble iron sulfide and pyrite. Oxygen diffusing downward reacts with the iron and oxides it, forming iron hydroxides. It is also possible that anoxygenic photosynthetic iron-oxidizing bacteria thrive in this layer, but hitherto, this has not been demonstrated. Sulfate-reducing bacteria are obligate anaerobic microorganisms, although some can tolerate low levels of oxygen or are even capable of respiring it, although they are unable to live by aerobic respiration. For these reasons, it has been assumed that sulfate-reducing bacteria are mostly present in the permanent anoxic layers of the microbial mat. This is not the case. Sulfate-reducing bacteria occur throughout the microbial mat although the different groups show a distinct depth distribution, particularly with respect to their oxygen sensitivity. Microbial mats have also been termed laminated microbial ecosystems. The lamination is the result of the vertically stratified communities of different functional groups of phototrophic microorganisms, notably the cyanobacteria and the purple sulfur bacteria (and in some cases also green sulfur bacteria), giving the mat a varicolored appearance. This lamination is also called an instantaneous biological lamination to distinguish it from a historical lamination which is produced by successive microbial mats, while the older mats are only partly degraded (compare with the rings of a tree). This is often seen under extreme conditions such as hypersaline or hyperthermal environments. This historical lamination is also preserved when the mat turns to stone, i.e., when calcification takes place.

Stromatolites are by definition laminated rock and are known from the fossil record up to almost 3.5 billion years before present. The majority of modern microbial mats do not turn to stone, although it is uncertain whether this has always been like this. The fossil record also indicates the presence of noncalcified microbial mats through features known as Microbial Induced Sediment Structures (MISS). It is not fully clear why some mats calcify and others apparently do not. It is hypothesized that Extracellular Polymeric Substances (EPS) produced by the Cyanobacteria bind calcium, thereby lowering the solubility product of calcium carbonate. When the EPS is degraded, calcium is locally released causing saturation of calcite. Modern marine stromatolites are rare and found only in the Exuma Cays (Bahamas) and in Shark Bay (Western Australia). Stromatolites with remarkable similarity to those of the Precambrian are found in alkaline environments which may hint to an alkaline Archean ocean. It has been shown that calcification occurs in microbial mats that do not lithify, but the calcite in these mats dissolves as a result of microbial metabolism.

References

- Chang, kenneth (13 april 2017). “conditions for life detected on saturn moon enceladus”. New york times. Retrieved 14 april 2017
- Biosphere, encyclopedia: nationalgeographic.org, Retrieved 3 May, 2019
- “Age of the earth”. U.s. geological survey. 1997. Archived from the original on 23 december 2005. Retrieved 2006-01-10
- Life-on-earth, life, science: britannica.com, Retrieved 5 February, 2019
- Pinti, daniele l.; arndt, nicholas (2014), “oceans, origin of”, encyclopedia of astrobiology, springer berlin heidelberg, pp. 1–5, doi:10.1007/978-3-642-27833-4-1098-4, isbn 9783642278334
- Significance-of-water-in-biosphere, water: yourarticlelibrary.com, Retrieved 30 April, 2019
- “Anu - research school of earth sciences - anu college of science - harrison”. Ses.anu.edu.au. Archived from the original on 2006-06-21. Retrieved 2009-08-20
- Biogenic-substance: wikivisually.com, Retrieved 15 July, 2019
- “Spacecraft data suggest saturn moon›s ocean may harbor hydrothermal activity”. Nasa. 11 march 2015. Retrieved 12 march 2015

3

Biomes

Biomes are the major communities, which are classified on the basis of predominant vegetation and characterized by the adaptation of organisms to that specific environment. Some of the different types of biomes are marine biomes, human micro biomes, endolithic biomes, anthropogenic biomes, etc. This chapter discusses in detail these types of biomes.

BIOME

Biomes are a way to divide the Earth's surface. These divisions are based on climate patterns, soil types, and the animals and plants that inhabit an area. There are biomes on dry land and in water. Every inch of the Earth's surface is a part of one or more biomes.

There are a couple of different ways to look at the number of biomes. Some like to divide biomes into five basic types: aquatic, forest, desert, tundra, and grassland. These five types of biomes can be further divided by differences in seasons or animal and plant species.

Aquatic Biome

The aquatic biome consists of any part of Earth that is covered with water. This includes freshwater and salt water. The aquatic biome can be further divided into freshwater biomes, marine biomes, wetland biomes, coral reef biomes, and estuaries. These subdivisions are based on the salt content of the water, the aquatic plants that live there, and the aquatic animals that thrive there.

Forest Biome

The forest biome is the largest and has a wide variety of plants, trees, animals, insects, and microscopic organisms. The major characteristic of the forest biome is its trees. About 30% of the Earth is considered a part of the forest biome. The forest biome is subdivided by its climate and types of trees present. These subdivisions are: the rainforest biome, temperate biome, chaparral biome, alpine biome, and taiga biome.

Desert Biome

The desert biome has one major, distinguishing characteristic, the fact that it has very little vegetation. The climate is rather extreme depending on its location. The deserts of Africa are extremely hot during the winters and warm throughout the rest of the year. There are also cold deserts such as those in Antarctica. These deserts are extremely cold during the winter and cold during the other seasons.

Tundra Biome

Tundra biomes are the coldest places on Earth. They are similar to a cold desert except they receive less rainfall and contain different animals and plants. Even though the conditions are harsh, the tundra biome does have plant and animal habitats.

Grassland Biome

The grassland biome is made of rolling hills of various grasses. They receive just enough rain to sustain grass but not enough to grow many trees. There are a few trees that will grow in grasslands but sporadic wildfires keep them under control. There are two types of grassland, the savannas and temperate grasslands.

HUMAN MICROBIOME

The human microbiome is the aggregate of all microbiota that reside on or within human tissues and biofluids along with the corresponding anatomical sites in which they reside, including the skin, mammary glands, placenta, seminal fluid, uterus, ovarian follicles, lung, saliva, oral mucosa, conjunctiva, biliary tract, and gastrointestinal tract. Types of human microbiota include bacteria, archaea, fungi, protists and viruses. Though micro-animals can also live on the human body, they are typically excluded from this definition. In the context of genomics, the term *human microbiome* is sometimes used to refer to the collective genomes of resident microorganisms; however, the term *human metagenome* has the same meaning.

Humans are colonized by many microorganisms; the traditional estimate is that the average human body is inhabited by ten times as many non-human cells as human cells, but more recent estimates have lowered that ratio to 3:1 or even to approximately the same number. Some microorganisms that colonize humans are commensal, meaning they co-exist without harming humans; others have a mutualistic relationship with their human hosts. Conversely, some non-pathogenic microorganisms can harm human hosts via the metabolites they produce, like trimethylamine, which the human body converts to trimethylamine N-oxide via FMO3-mediated oxidation. Certain microorganisms perform tasks that are known to be useful to the human host but the role of most of them is not well understood. Those that are expected to be present, and that under normal circumstances do not cause disease, are sometimes deemed normal flora or normal microbiota.

Types

Bacteria

Populations of microbes (such as bacteria and yeasts) inhabit the skin and mucosal surfaces in various parts of the body. Their role forms part of normal, healthy human physiology, however if microbe numbers grow beyond their typical ranges (often due to a compromised immune system) or if microbes populate (such as through poor hygiene or injury) areas of the body normally not colonized or sterile (such as the blood, or the lower respiratory tract, or the abdominal cavity), disease can result (causing, respectively, bacteremia/sepsis, pneumonia, and peritonitis).

The Human Microbiome Project found that individuals host thousands of bacterial types, different body sites having their own distinctive communities. Skin and vaginal sites showed smaller diversity than the mouth and gut, these showing the greatest richness. The bacterial makeup for a given site on a body varies from person to person, not only in type, but also in abundance. Bacteria of the same species found throughout the mouth are of multiple subtypes, preferring to inhabit distinctly different locations in the mouth. Even the enterotypes in the human gut, previously thought to be well understood, are from a broad spectrum of communities with blurred taxon boundaries.

It is estimated that 500 to 1,000 species of bacteria live in the human gut but belong to just a few phyla: Firmicutes and Bacteroidetes dominate but there are also Proteobacteria, Verrumicrobia, Actinobacteria, Fusobacteria and Cyanobacteria.

A number of types of bacteria, such as *Actinomyces viscosus* and *A. naeslundii*, live in the mouth, where they are part of a sticky substance called plaque. If this is not removed by brushing, it hardens into calculus (also called tartar). The same bacteria also secrete acids that dissolve tooth enamel, causing tooth decay.

The vaginal microflora consist mostly of various lactobacillus species. It was long thought that the most common of these species was *Lactobacillus acidophilus*, but it has later been shown that *L. iners* is in fact most common, followed by *L. crispatus*. Other lactobacilli found in the vagina are *L. jensenii, L. delbruekii* and *L. gasseri*. Disturbance of the vaginal flora can lead to infections such as bacterial vaginosis or candidiasis ("yeast infection").

Archaea

Archaea are present in the human gut, but, in contrast to the enormous variety of bacteria in this organ, the numbers of archaeal species are much more limited. The dominant group are the methanogens, particularly *Methanobrevibacter smithii* and *Methanosphaera stadtmanae*. However, colonization by methanogens is variable, and only about 50% of humans have easily detectable populations of these organisms.

As of 2007, no clear examples of archaeal pathogens were known, although a relationship has been proposed between the presence of some methanogens and human periodontal disease.

Fungi

Fungi, in particular yeasts, are present in the human gut. The best-studied of these are *Candida* species due to their ability to become pathogenic in immunocompromised and even in healthy hosts. Yeasts are also present on the skin, such as *Malassezia* species, where they consume oils secreted from the sebaceous glands.

Viruses

Viruses, especially bacterial viruses (bacteriophages), colonize various body sites. These colonized sites include the skin, gut, lungs, and oral cavity. Virus communities have been associated with some diseases, and do not simply reflect the bacterial communities.

Anatomical areas

Skin

A study of twenty skin sites on each of ten healthy humans found 205 identified genera in nineteen bacterial phyla, with most sequences assigned to four phyla: Actinobacteria (51.8%), Firmicutes (24.4%), Proteobacteria (16.5%), and Bacteroidetes (6.3%). A large number of fungal genera are present on healthy human skin, with some variability by region of the body; however, during pathological conditions, certain genera tend to dominate in the affected region. For example, *Malassezia* is dominant in atopic dermatitis and *Acremonium* is dominant on dandruff-afflicted scalps.

The skin acts as a barrier to deter the invasion of pathogenic microbes. The human skin contains microbes that reside either in or on the skin and can be residential or transient. Resident microorganism types vary in relation to skin type on the human body. A majority of microbes reside on superficial cells on the skin or prefer to associate with glands. These glands such as oil or sweat glands provide the microbes with water, amino acids, and fatty acids. In addition, resident bacteria that associated with oil glands are often Gram-positive and can be pathogenic.

Conjunctiva

A small number of bacteria and fungi are normally present in the conjunctiva. Classes of bacteria include Gram-positive cocci (e.g., *Staphylococcus* and Streptococcus) and

Gram-negative rods and cocci (e.g., *Haemophilus* and *Neisseria*) are present. Fungal genera include *Candida*, *Aspergillus*, and *Penicillium*. The lachrymal glands continuously secrete, keeping the conjunctiva moist, while intermittent blinking lubricates the conjunctiva and washes away foreign material. Tears contain bactericides such as lysozyme, so that microorganisms have difficulty in surviving the lysozyme and settling on the epithelial surfaces.

Gastrointestinal Tract

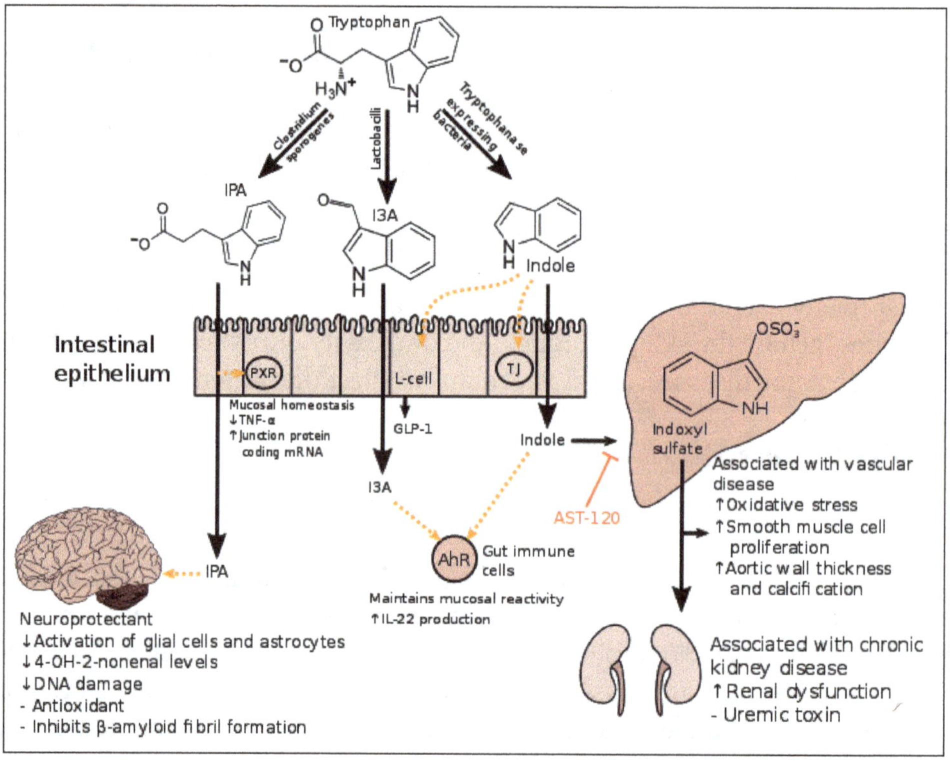

The above figure shows the biosynthesis of bioactive compounds (indole and certain other derivatives) from tryptophan by bacteria in the gut. Indole is produced from tryptophan by bacteria that express tryptophanase. Clostridium sporogenes metabolizes tryptophan into indole and subsequently 3-indolepropionic acid (IPA), a highly potent neuroprotective antioxidant that scavenges hydroxyl radicals. IPA binds to the pregnane X receptor (PXR) in intestinal cells, thereby facilitating mucosal homeostasis and barrier function. Following absorption from the intestine and distribution to the brain, IPA confers a neuroprotective effect against cerebral ischemia and Alzheimer's disease. Lactobacillus species metabolize tryptophan into indole-3-aldehyde (I3A) which acts on the aryl hydrocarbon receptor (AhR) in intestinal immune cells, in turn increasing interleukin-22 (IL-22) production. Indole itself triggers the secretion of glucagon-like peptide-1 (GLP-1) in intestinal L cells and acts as a ligand for AhR. Indole can also be metabolized by the liver into indoxyl sulfate, a compound that is toxic in high concentrations and associated with vascular disease and renal dysfunction. AST-120 (activated charcoal), an intestinal sorbent that is taken by mouth, adsorbs indole, in turn decreasing the concentration of indoxyl sulfate in blood plasma.

In humans, the composition of the gastrointestinal microbiome is established during birth. Birth by Cesarean section or vaginal delivery also influences the gut's microbial composition. Babies born through the vaginal canal have non-pathogenic, beneficial gut microbiota similar to those found in the mother. However, the gut microbiota of babies delivered by C-section harbors more pathogenic bacteria such as Escherichia coli and Staphylococcus and it takes longer to develop non-pathogenic, beneficial gut microbiota.

The relationship between some gut flora and humans is not merely commensal (a non-harmful coexistence), but rather a mutualistic relationship. Some human gut microorganisms benefit the host by fermenting dietary fiber into Short-Chain Fatty Acids (SCFAs), such as acetic acid and butyric acid, which are then absorbed by the host. Intestinal bacteria also play a role in synthesizing vitamin B and vitamin K as well as metabolizing bile acids, sterols, and xenobiotics. The systemic importance of the SCFAs and other compounds they produce are like hormones and the gut flora itself appears to function like an endocrine organ, and dysregulation of the gut flora has been correlated with a host of inflammatory and autoimmune conditions.

The composition of human gut flora changes over time, when the diet changes, and as overall health changes. A systematic review of 15 human randomized controlled trials from July 2016 found that certain commercially available strains of probiotic bacteria from the *Bifidobacterium* and *Lactobacillus* genera (*B. longum, B. breve, B. infantis, L. helveticus, L. rhamnosus, L. plantarum*, and *L. casei*), when taken by mouth in daily doses of 10^9–10^{10} Colony Forming Units (CFU) for 1–2 months, possess treatment efficacy (i.e., improves behavioral outcomes) in certain central nervous system disorders – including anxiety, depression, autism spectrum disorder, and obsessive–compulsive disorder – and improves certain aspects of memory. However, changes in the composition of gut microbiota has also been found to be correlated with harmful effects on health.Of the microbiota that causes obesity rather than the other way around. This was done by transplanting the gut microbiota from Diet-Induced Obese (DIO) mice or lean control mice into lean germ-free mice that do not have a microbiome. They found that the mice transplanted with DIO mouse gut microbiota had significantly higher total body fat than the mice transplanted with lean mouse microbiota when fed with the same diet.

Urethra and Bladder

The genitourinary system appears to have a microbiota, which is an unexpected finding in light of the long-standing use of standard clinical microbiological culture methods to detect bacteria in urine when people show signs of a urinary tract infection; it is common for these tests to show no bacteria present. It appears that common culture methods do not detect many kinds of bacteria and other microorganisms that are normally present. As of 2017, sequencing methods were used to identify these microorganisms to determine if there are differences in microbiota between people with urinary tract problems and those who are healthy. To properly assess the microbiome of the bladder as opposed to the genitourinary system, the urine specimen should be collected directly from the bladder, which is often done with a catheter.

Vagina

Vaginal microbiota refers to those species and genera that colonize the vagina. These organisms play an important role in protecting against infections and maintaining vaginal health. The most abundant vaginal microorganisms found in premenopausal women are from the genus *Lactobacillus*, which suppress pathogens by producing hydrogen peroxide and lactic acid. Bacterial species composition and ratios vary depending on the stage of the menstrual cycle. Ethnicity also influences vaginal flora. The occurrence of hydrogen peroxide-producing lactobacilli is lower in African American women and vaginal pH is higher. Other influential factors such as sexual intercourse and antibiotics have been linked to the loss of lactobacilli. Moreover, studies have found that sexual intercourse with a condom does appear to change lactobacilli levels, and does increase the level of *Escherichia coli* within the vaginal flora. Changes in the normal, healthy vaginal microbiota is an indication of infections, such as candidiasis or bacterial vaginosis. *Candida albicans* inhibits the growth of *Lactobacillus* species, while *Lactobacillus* species which produce hydrogen peroxide inhibit the growth and virulence of *Candida albicans* in both the vagina and the gut.

Fungal genera that have been detected in the vagina include *Candida, Pichia, Eurotium, Alternaria, Rhodotorula,* and *Cladosporium*, among others.

Placenta

Until recently the placenta was considered to be a sterile organ but commensal, nonpathogenic bacterial species and genera have been identified that reside in the placental tissue.

Uterus

Until recently, the upper reproductive tract of women was considered to be a sterile environment. A variety of microorganisms inhabit the uterus of healthy, asymptomatic women of reproductive age. The microbiome of the uterus differs significantly from that of the vagina and gastrointestinal tract.

Oral Cavity

The environment present in the human mouth allows the growth of characteristic microorganisms found there. It provides a source of water and nutrients, as well as a moderate temperature. Resident microbes of the mouth adhere to the teeth and gums to resist mechanical flushing from the mouth to stomach where acid-sensitive microbes are destroyed by hydrochloric acid.

Anaerobic bacteria in the oral cavity include: *Actinomyces, Arachnia, Bacteroides, Bifidobacterium, Eubacterium, Fusobacterium, Lactobacillus, Leptotrichia, Peptococcus, Peptostreptococcus, Propionibacterium, Selenomonas, Treponema*, and *Veillonella*. Genera of fungi that are frequently found in the mouth include *Candida, Cladosporium, Aspergillus, Fusarium, Glomus, Alternaria, Penicillium*, and *Cryptococcus*, among others.

Bacteria accumulate on both the hard and soft oral tissues in biofilm allowing them to adhere and strive in the oral environment while protected from the environmental factors and antimicrobial agents. Saliva plays a key biofilm homeostatic role allowing recolonization of bacteria for formation and controlling growth by detaching biofilm buildup. It also provides a means of nutrients and temperature regulation. The location of the biofilm determines the type of exposed nutrients it receives.

Oral bacteria have evolved mechanisms to sense their environment and evade or modify the host. However, a highly efficient innate host defense system constantly monitors the bacterial colonization and prevents bacterial invasion of local tissues. A dynamic equilibrium exists between dental plaque bacteria and the innate host defense system.

This dynamic between host oral cavity and oral microbes plays a key role in health and disease as it provides entry into the body. A healthy equilibrium presents a symbiotic relationship where oral microbes limit growth and adherence of pathogens while the host provides an environment for them to flourish. Ecological changes such as change of immune status, shift of resident microbes and nutrient availability shift from a mutual to parasitic relationship resulting in the host being prone to oral and systemic disease. Systemic diseases such as diabetes and cardiovascular diseases has been correlated to poor oral health. Of particular interest is the role of oral microorganisms in the two major dental diseases: dental caries and periodontal disease. Pathogen colonization at the periodontium cause an excessive immune response resulting in a periodontal pocket- a deepened space between the tooth and gingiva. This acts as a protected blood-rich reservoir with nutrients for anaerobic pathogens. Systemic disease at various sites of the body can result from oral microbes entering the blood bypassing periodontal pockets and oral membranes.

Persistent proper oral hygiene is the primary method for preventing oral and systemic disease. It reduces the density of biofilm and overgrowth of potential pathogenic bacteria resulting in disease. However, proper oral hygiene may not be enough as the oral microbiome, genetics, and changes to immune response play a factor in developing chronic infections. Use of antibiotics could treat already spreading infection but ineffective against bacteria within biofilms.

Lung

Much like the oral cavity, the upper and lower respiratory system possess mechanical deterrents to remove microbes. Goblet cells produce mucous which traps microbes and moves them out of the respiratory system via continuously moving ciliated epithelial cells. In addition, a bactericidal effect is generated by nasal mucus which contains the enzyme lysozyme. The upper and lower respiratory tract appears to have its own set of microbiota. Pulmonary bacterial microbiota belong to 9 major bacterial genera: *Prevotella*, *Sphingomonas*, *Pseudomonas*, *Acinetobacter*, *Fusobacterium*, *Megasphaera*, *Veillonella*, *Staphylococcus*, and *Streptococcus*. Some of the bacteria considered "normal biota" in the respiratory tract can cause serious disease especially in immunocompromised individuals; these include *Streptococcus pyogenes*, *Haemophilus influenzae*, *Streptococcus pneumoniae*, *Neisseria meningitidis*, and *Staphylococcus aureus*. Fungal genera that compose the pulmonary mycobiome include *Candida*, *Malassezia*, *Neosartorya*, *Saccharomyces*, and *Aspergillus*, among others.

Unusual distributions of bacterial and fungal genera in the respiratory tract is observed in people with cystic fibrosis. Their bacterial flora often contains antibiotic-resistant and slow-growing bacteria, and the frequency of these pathogens changes in relation to age.

Biliary Tract

Traditionally the biliary tract has been considered to be normally sterile, and the presence of microorganisms in bile is a marker of pathological process. This assumption was confirmed by failure in allocation of bacterial strains from the normal bile duct. Papers began emerging in 2013 showing that the normal biliary microbiota is a separate functional layer which protects a biliary tract from colonization by exogenous microorganisms.

Disease and Death

A symbiotic relationship between the gut microbiota and different bacteria may influence an individual's immune response.

Cancer

Although cancer is generally a disease of host genetics and environmental factors, microorganisms are implicated in some 20% of human cancers. Particularly for potential factors in colon cancer, bacterial density is one million times higher than in the small intestine, and approximately 12-fold more cancers occur in the colon compared to the small intestine, possibly establishing a pathogenic role for microbiota in colon and rectal cancers. Microbial density may be used as a prognostic tool in assessment of colorectal cancers.

The microbiota may affect carcinogenesis in three broad ways: (i) altering the balance of tumor cell proliferation and death, (ii) regulating immune system function, and (iii) influencing metabolism of host-produced factors, foods and pharmaceuticals. Tumors arising at boundary surfaces, such as the skin, oropharynx and respiratory, digestive and urogenital tracts, harbor a microbiota. Substantial microbe presence at a tumor site does not establish association or causal links. Instead, microbes may find tumor oxygen tension or nutrient profile supportive. Decreased populations of specific microbes or induced oxidative stress may also increase risks. Of the around 10^{30} microbes on earth, ten are designated by the International Agency for Research on Cancer as human carcinogens. Microbes may secrete proteins or other factors directly drive cell proliferation in the host, or may up- or down-regulate the host immune system including driving acute or chronic inflammation in ways that contribute to carcinogenesis.

Concerning the relationship of immune function and development of inflammation, mucosal surface barriers are subject to environmental risks and must rapidly repair to maintain homeostasis. Compromised host or microbiota resiliency also reduce resistance to malignancy, possibly inducing inflammation and cancer. Once barriers are breached, microbes can elicit proinflammatory or immunosuppressive programs through various pathways. For example, cancer-associated microbes appear to activate NF-κB signaling within the tumor microenviroment. Other pattern recognition receptors, such as nucleotide-binding oligomerization domain–like receptor (NLR) family members *NOD-2*, *NLRP3*, *NLRP6* and *NLRP12*, may play a role in mediating colorectal cancer. Likewise *Helicobacter pylori* appears to increase the risk of gastric cancer, due to its driving a chronic inflammatory response in the stomach.

Inflammatory Bowel Disease

Inflammatory bowel disease consists of two different diseases: ulcerative colitis and Crohn's disease and both of these diseases present with disruptions in the gut microbiota (also known as dysbiosis). This dysbiosis presents itself in the form of decreased microbial diversity in the gut, and is correlated to defects in host genes that changes the innate immune response in individuals.

Human Immunodeficiency Virus

The HIV disease progression influences the composition and function of the gut microbiota, with notable differences between HIV-negative, HIV-positive, and post-ART HIV-positive populations. HIV decreases the integrity of the gut epithelial barrier function by affecting tight junctions. This breakdown allows for translocation across the gut epithelium, which is thought to contribute to increases in inflammation seen in people with HIV.

Vaginal microbiota plays a role in the infectivity of HIV, with an increased risk of infection and transmission when the woman has bacterial vaginosis, a condition characterized by an abnormal balance of vaginal bacteria. The enhanced infectivity is seen with the increase in pro-inflammatory cytokines and CCR5+CD4+ cells in the vagina. However, a decrease in infectivity is seen with increased levels of vaginal *Lactobacillus,* which promotes an anti-inflammatory condition.

Death

With death, the microbiome of the living body collapses and a different composition of microorganisms named *necrobiome* establishes itself in order to help with the complex decomposition process. Its predictable changes over time are thought to be useful to help determine the time of death.

Migration

Immediate changes in the microbiota may occur when a person migrates from one country to another, such as when Thai immigrants settled in the United States. Losses of microbiota diversity were greater in obese individuals and children of immigrants.

BIOGEOGRAPHIC REALMS

A biogeographic realm or ecozone is the broadest biogeographic division of the Earth's land surface, based on distributional patterns of terrestrial organisms. They are subdivided in ecoregions, which are classified in biomes or habitat types.

The realms delineate the large areas of the Earth's surface within which organisms have been evolving in relative isolation over long periods of time, separated from one another by geographic features, such as oceans, broad deserts, or high mountain ranges, that constitute barriers to migration. As such, biogeographic realms designations are used to indicate general groupings of organisms based on their shared biogeography. Biogeographic realms correspond to the floristic kingdoms of botany or zoogeographic regions of zoology.

Biogeographic realms are characterized by the evolutionary history of the organisms they contain. They are distinct from biomes, also known as major habitat types, which are divisions of the Earth's surface based on *life form*, or the adaptation of animals, fungi, micro-organisms and plants to climatic, soil, and other conditions. Biomes are characterized by similar climax vegetation. Each realm may include a number of different biomes. A tropical moist broadleaf forest in Central America, for example, may be similar to one in New Guinea in its vegetation type and structure, climate, soils, etc., but these forests are inhabited by animals, fungi, micro-organisms and plants with very different evolutionary histories.

The patterns of distribution of living organisms in the world's biogeographic realms were shaped by the process of plate tectonics, which has redistributed the world's land masses over geological history.

Terrestrial Biogeographic Realms

The World Wildlife Fund scheme is broadly similar to Miklos Udvardy's system, the chief difference being the delineation of the Australasian realm relative to the Antarctic, Oceanic, and Indomalayan realms. In the WWF system, The Australasia realm includes Australia, Tasmania, the islands of Wallacea, New Guinea, the East Melanesian islands, New Caledonia, and New Zealand. Udvardy's Australian realm includes only Australia and Tasmania; he places Wallacea in the Indomalayan Realm, New Guinea, New Caledonia, and East Melanesia in the Oceanian Realm, and New Zealand in the Antarctic Realm.

Biogeographic realm	Area		Notes
	million square kilometres	million square miles	
Palearctic	54.1	20.9	Including the bulk of Eurasia and North Africa.
Nearctic	22.9	8.8	Including most of North America.
Afrotropic	22.1	8.5	Including Trans-Saharan Africa and Arabia.
Neotropic	19.0	7.3	Including South America, Central America, and the Caribbean.
Australasia	7.6	2.9	Including Australia, New Guinea, New Zealand, and neighbouring islands. The northern boundary of this zone is known as the Wallace line.
Indo-Malaya	7.5	2.9	Including the Indian subcontinent, Southeast Asia, and southern China.
Oceania	1.0	0.39	Including Polynesia (except New Zealand), Micronesia, and the Fijian Islands.
Antarctic	0.3	0.12	Including Antarctica.

The Palearctic and Nearctic are sometimes grouped into the Holarctic realm.

Morrone Biogeographic Kingdoms

Following the nomenclatural conventions set out in the International Code of Area Nomenclature, Morrone defined the next biogeographic kingdoms (or realms) and regions:

- Holarctic kingdom Heilprin:
    - Nearctic region Sclater.

    - Palearctic region Sclater.
- Holotropical kingdom Rapoport:
    - Neotropical region Sclater.
    - Ethiopian region Sclater.
    - Oriental region Wallace.
- Austral kingdom Engler:
    - Cape region Grisebach.
    - Andean region Engler.
    - Australian region Sclater.
    - Antarctic region Grisebach.
- Transition zones:
    - Mexican transition zone (Nearctic–Neotropical transition).
    - Saharo-Arabian transition zone (Palearctic–Ethiopian transition).
    - Chinese transition zone (Palearctic–Oriental transition zone transition).
    - Indo-Malayan, Indonesian or Wallace's transition zone (Oriental–Australian transition).
    - South American transition zone (Neotropical–Andean transition).

Freshwater Biogeographic Realms

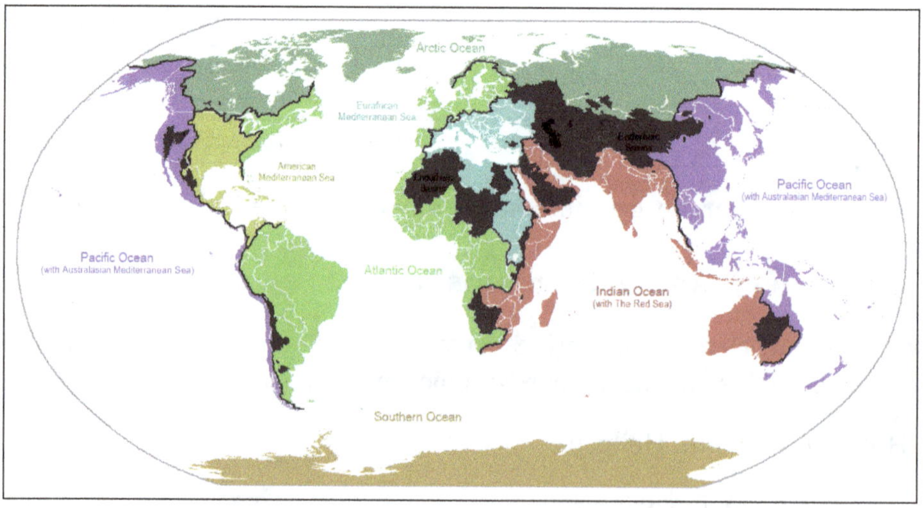

The applicability of Udvardy scheme to most freshwater taxa is unresolved.

The drainage basins of the principal oceans and seas of the world are marked by continental divides. The grey areas are endorheic basins that do not drain to the ocean.

ECOREGION

An ecoregion (ecological region) is an ecologically and geographically defined area that is smaller than a bioregion, which in turn is smaller than an ecozone. All three of these are either less or greater than an ecosystem. Ecoregions cover relatively large areas of land or water, and contain characteristic, geographically distinct assemblages of natural communities and species. The biodiversity of flora, fauna and ecosystems that characterise an ecoregion tends to be distinct from that of other ecoregions. In theory, biodiversity or conservation ecoregions are relatively large areas of land or water where the probability of encountering different species and communities at any given point remains relatively constant, within an acceptable range of variation (largely undefined at this point).

Three caveats are appropriate for all bio-geographic mapping approaches. Firstly, no single bio-geographic framework is optimal for all taxa. Ecoregions reflect the best compromise for as many taxa as possible. Secondly, ecoregion boundaries rarely form abrupt edges; rather, ecotones and mosaic habitats bound them. Thirdly, most ecoregions contain habitats that differ from their assigned biome. Biogeographic provinces may originate due to various barriers. Some physical (plate tectonics, topographic highs), some climatic (latitudinal variation, seasonal range) and some ocean chemical related (salinity, oxygen levels).

The Ötztal Alps, a mountain range in the central Alps of Europe, are part of the Central Eastern Alps, and can both be termed as ecoregions.

An ecoregion is a "recurring pattern of ecosystems associated with characteristic combinations of soil and landform that characterise that region". Omernik elaborates on this by defining ecoregions as: "areas within which there is spatial coincidence in characteristics of geographical phenomena associated with differences in the quality, health, and integrity of ecosystems". "Characteristics of geographical phenomena" may include geology, physiography, vegetation, climate, hydrology, terrestrial and aquatic fauna, and soils, and may or may not include the impacts of human activity (e.g. land use patterns, vegetation changes). There is significant, but not absolute, spatial correlation among these characteristics, making the delineation of ecoregions an imperfect science. Another complication is that environmental conditions across an ecoregion boundary may change very gradually, e.g. the prairie-forest transition in the midwestern United States, making it difficult to identify an exact dividing boundary. Such transition zones are called ecotones.

A conifer forest in the Swiss Alps (National Park).

Ecoregions can be categorized using an algorithmic approach or a holistic, "weight-of-evidence" approach where the importance of various factors may vary. An example of the algorithmic approach is Robert Bailey's work for the U.S. Forest Service, which uses a hierarchical classification that first divides land areas into very large regions based on climatic factors, and subdivides these regions, based first on dominant potential vegetation, and then by geomorphology and soil characteristics. The weight-of-evidence approach is exemplified by James Omernik's work for the United States Environmental Protection Agency, subsequently adopted (with modification) for North America by the Commission for Environmental Cooperation.

The intended purpose of ecoregion delineation may affect the method used. For example, the WWF ecoregions were developed to aid in biodiversity conservation planning, and place a greater emphasis than the Omernik or Bailey systems on floral and faunal differences between regions. The WWF classification defines an ecoregion as given below.

A large area of land or water that contains a geographically distinct assemblage of natural communities that:

- Share a large majority of their species and ecological dynamics;

- Share similar environmental conditions, and;
- Interact ecologically in ways that are critical for their long-term persistence.

According to WWF, the boundaries of an ecoregion approximate the original extent of the natural communities prior to any major recent disruptions or changes. WWF has identified 867 terrestrial ecoregions, and approximately 450 freshwater ecoregions across the Earth.

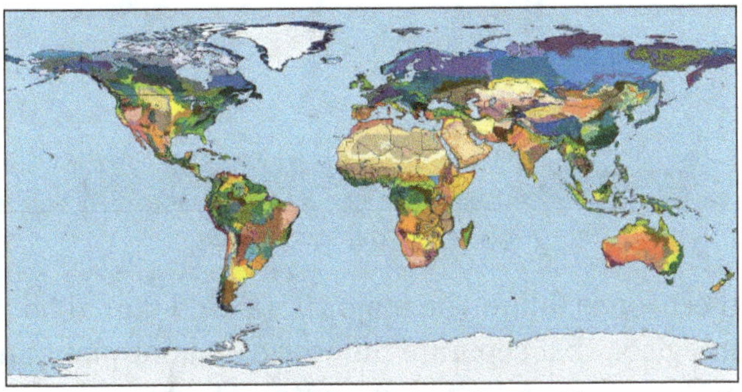

Terrestrial Ecoregions of the World.

Importance

The use of the term ecoregion is an outgrowth of a surge of interest in ecosystems and their functioning. In particular, there is awareness of issues relating to spatial scale in the study and management of landscapes. It is widely recognized that interlinked ecosystems combine to form a whole that is "greater than the sum of its parts". There are many attempts to respond to ecosystems in an integrated way to achieve "multi-functional" landscapes, and various interest groups from agricultural researchers to conservationists are using the "ecoregion" as a unit of analysis.

The "Global 200" is the list of ecoregions identified by WWF as priorities for conservation.

Ecologically based movements like bioregionalism maintain that ecoregions, rather than arbitrarily defined political boundaries, provide a better foundation for the formation and governance of human communities, and have proposed ecoregions and watersheds as the basis for bioregional democracy initiatives.

Terrestrial

Terrestrial ecoregions are land ecoregions, as distinct from freshwater and marine ecoregions. In this context, *terrestrial* is used to mean "of land" (soil and rock), rather than the more general sense "of Earth" (which includes land and oceans).

WWF (World Wildlife Fund) ecologists currently divide the land surface of the Earth into 8 major ecozones containing 867 smaller terrestrial ecoregions. The WWF effort is a synthesis of many previous efforts to define and classify ecoregions. Many consider this classification to be quite decisive, and some propose these as stable borders for bioregional democracy initiatives.

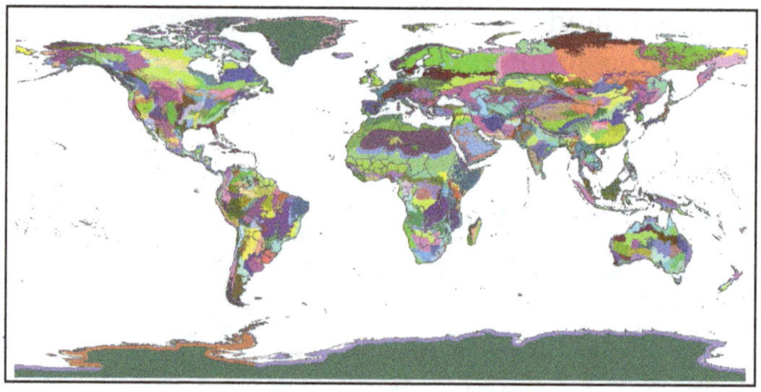

WWF terrestrial ecoregions.

The eight terrestrial ecozones follow the major floral and faunal boundaries, identified by botanists and zoologists, that separate the world's major plant and animal communities. Ecozone boundaries generally follow continental boundaries, or major barriers to plant and animal distribution, like the Himalayas and the Sahara. The boundaries of ecoregions are often not as decisive or well recognized, and are subject to greater disagreement.

Ecoregions are classified by biome type, which are the major global plant communities determined by rainfall and climate. Forests, grasslands (including savanna and shrubland), and deserts (including xeric shrublands) are distinguished by climate (tropical and subtropical vs. temperate and boreal climates) and, for forests, by whether the trees are predominantly conifers (gymnosperms), or whether they are predominantly broadleaf (Angiosperms) and mixed (broadleaf and conifer). Biome types like Mediterranean forests, woodlands, and scrub; tundra; and mangroves host very distinct ecological communities, and are recognized as distinct biome types as well.

Marine

View of Earth, taken in 1972 by the Apollo 17 crew. Approximately 72% of the Earth's surface (an area of some 361 million square kilometers) consists of ocean.

Marine ecoregions are: "Areas of relatively homogeneous species composition, clearly distinct from adjacent systems. In ecological terms, these are strongly cohesive units, sufficiently large to encompass ecological or life history processes for most sedentary species." They have been defined by The Nature Conservancy (TNC) and World Wildlife Fund (WWF) to aid in conservation activities for marine ecosystems. Forty-three priority marine ecoregions were delineated as part of WWF's Global 200 efforts. The scheme used to designate and classify marine ecoregions is analogous to that used for terrestrial ecoregions. Major habitat types are identified: polar, temperate shelves and seas, temperate upwelling, tropical upwelling, tropical coral, pelagic (trades and westerlies), abyssal, and hadal (ocean trench). These correspond to the terrestrial biomes.

The Global 200 classification of marine ecoregions is not developed to the same level of detail and comprehensiveness as that of the terrestrial ecoregions; only the priority conservation areas are listed.

The 232 individual marine ecoregions are grouped into 62 marine provinces, which in turn group into 12 marine realms, which represent the broad latitudinal divisions of polar, temperate, and tropical seas, with subdivisions based on ocean basins (except for the southern hemisphere temperate oceans, which are based on continents).

Major biogeographic realms, analogous to the eight terrestrial ecozones, represent large regions of the ocean basins: Arctic, Temperate Northern Atlantic, Temperate Northern Pacific, Tropical Atlantic, Western Indo-Pacific, Central Indo-Pacific, Eastern Indo-Pacific, Tropical Eastern Pacific, Temperate South America, Temperate Southern Africa, Temperate Australasia, Southern Ocean.

A similar system of identifying areas of the oceans for conservation purposes is the system of large marine ecosystems (LMEs), developed by the US National Oceanic and Atmospheric Administration (NOAA).

Freshwater

A freshwater ecoregion is a large area encompassing one or more freshwater systems that contains a distinct assemblage of natural freshwater communities and species. The freshwater species, dynamics, and environmental conditions within a given ecoregion are more similar to each other than to those of surrounding ecoregions and together form a conservation unit. Freshwater systems include rivers, streams, lakes, and wetlands. Freshwater ecoregions are distinct from terrestrial ecoregions, which identify biotic communities of the land, and marine ecoregions, which are biotic communities of the oceans.

A map of Freshwater Ecoregions of the World, released in 2008, has 426 ecoregions covering virtually the entire non-marine surface of the earth.

World Wildlife Fund (WWF) identifies twelve major habitat types of freshwater ecoregions: Large lakes, large river deltas, polar freshwaters, montane freshwaters, temperate coastal rivers, temperate floodplain rivers and wetlands, temperate upland rivers, tropical and subtropical coastal rivers, tropical and subtropical floodplain rivers and wetlands, tropical and subtropical upland rivers, xeric freshwaters and endorheic basins, and oceanic islands. The freshwater major habitat types reflect groupings of ecoregions with similar biological, chemical, and physical characteristics and are roughly equivalent to biomes for terrestrial systems.

The Global 200, a set of ecoregions identified by WWF whose conservation would achieve the goal of saving a broad diversity of the Earth's ecosystems, includes a number of areas highlighted for their freshwater biodiversity values. The Global 200 preceded Freshwater Ecoregions of the World and incorporated information from regional freshwater ecoregional assessments that had been completed at that time.

MARINE BIOMES

Marine habitats are habitats that support marine life. Marine life depends in some way on the saltwater that is in the sea. A habitat is an ecological or environmental area inhabited by one or more living species. The marine environment supports many kinds of these habitats.

Marine habitats can be divided into coastal and open ocean habitats. Coastal habitats are found in the area that extends from as far as the tide comes in on the shoreline out to the edge of the continental shelf. Most marine life is found in coastal habitats, even though the shelf area occupies only seven percent of the total ocean area. Open ocean habitats are found in the deep ocean beyond the edge of the continental shelf.

Alternatively, marine habitats can be divided into pelagic and demersal zones. Pelagic habitats are found near the surface or in the open water column, away from the bottom of the ocean. Demersal habitats are near or on the bottom of the ocean. An organism living in a pelagic habitat is said to be a pelagic organism, as in pelagic fish. Similarly, an organism living in a demersal habitat is said to be a demersal organism, as in demersal fish. Pelagic habitats are intrinsically shifting and ephemeral, depending on what ocean currents are doing.

Marine habitats can be modified by their inhabitants. Some marine organisms, like corals, kelp, mangroves and seagrasses, are ecosystem engineers which reshape the marine environment to the point where they create further habitat for other organisms. By volume, oceans provide about 99 percent of the living space on the planet.

Ocean Currents

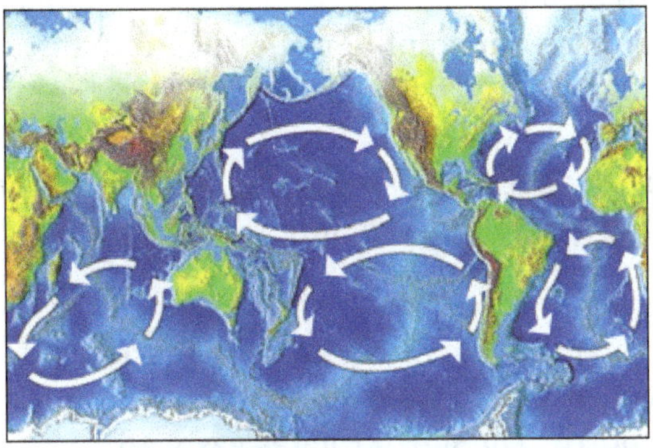

Ocean gyres rotate clockwise in the north and counterclockwise in the south.

In marine systems, ocean currents have a key role determining which areas are effective as habitats, since ocean currents transport the basic nutrients needed to support marine life. Plankton are the life forms that inhabit the ocean that are so small (less than 2 mm) that they cannot effectively propel themselves through the water, but must drift instead with the currents. If the current carries the right nutrients, and if it also flows at a suitably shallow depth where there is plenty of sunlight, then such a current itself can become a suitable habitat for photosynthesizing tiny algae called phytoplankton. These tiny plants are the primary producers in the ocean, at the start of the food chain. In turn, as the population of drifting phytoplankton grows, the water becomes a suitable habitat for zooplankton, which feed on the phytoplankton. While phytoplankton are tiny drifting plants, zooplankton are tiny drifting animals, such as the larvae of fish and marine invertebrates. If sufficient zooplankton establish themselves, the current becomes a candidate habitat for the forage fish that feed on them. And then if sufficient forage fish move to the area, it becomes a candidate habitat for larger predatory fish and other marine animals that feed on the forage fish. In this dynamic way, the current itself can, over time, become a moving habitat for multiple types of marine life.

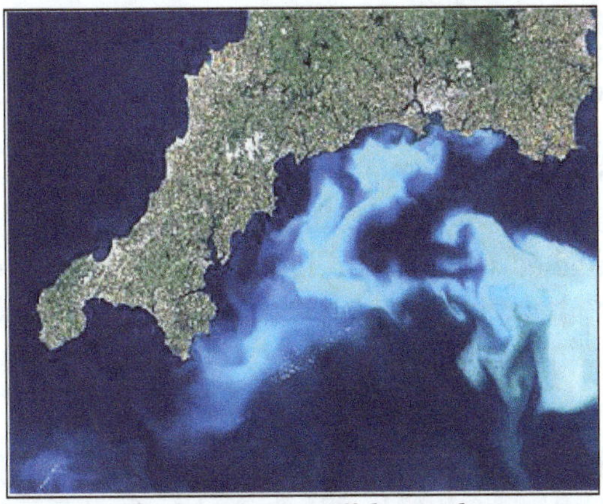

This algae bloom occupies sunlit epipelagic waters off the southern coast of England. The algae are maybe feeding on nutrients from land runoff or upwellings at the edge of the continental shelf.

Ocean currents can be generated by differences in the density of the water. How dense water is depends on how saline or warm it is. If water contains differences in salt content or temperature, then the different densities will initiate a current. Water that is saltier or cooler will be denser, and will sink in relation to the surrounding water. Conversely, warmer and less salty water will float to the surface. Atmospheric winds and pressure differences also produces surface currents, waves and seiches. Ocean currents are also generated by the gravitational pull of the sun and moon (tides), and seismic activity (tsunami).

The rotation of the Earth affects the direction ocean currents take, and explains which way the large circular ocean gyres rotate in the image above left. Suppose a current at the equator is heading north. The Earth rotates eastward, so the water possesses that rotational momentum. But the further the water moves north, the slower the earth moves eastward. If the current could get to the North Pole, the earth wouldn't be moving eastward at all. To conserve its rotational momentum, the further the current travels north the faster it must move eastward. So the effect is that the current curves to the right. This is the Coriolis effect. It is weakest at the equator and strongest at the poles. The effect is opposite south of the equator, where currents curve left.

Marine Topography

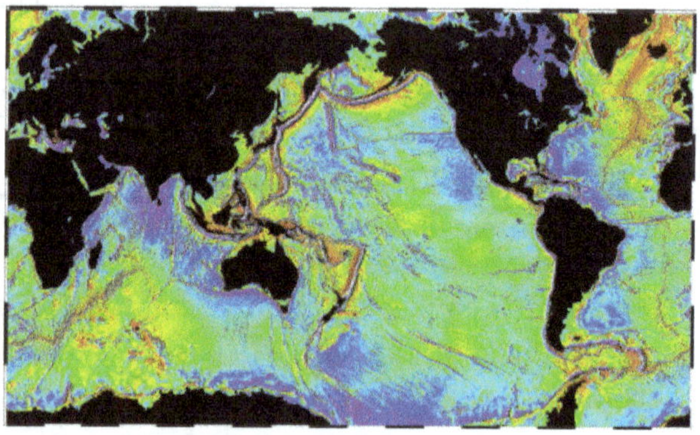

Map of underwater topography.

Marine (or seabed or ocean) topography refers to the shape the land has when it interfaces with the ocean. These shapes are obvious along coastlines, but they occur also in significant ways underwater. The effectiveness of marine habitats is partially defined by these shapes, including the way they interact with and shape ocean currents, and the way sunlight diminishes when these landforms occupy increasing depths. Tidal networks depend on the balance between sedimentary processes and hydrodynamics however, anthropogenic influences can impact the natural system more than any physical driver.

Marine topographies include coastal and oceanic landforms ranging from coastal estuaries and shorelines to continental shelves and coral reefs. Further out in the open ocean, they include underwater and deep sea features such as ocean rises and seamounts. The submerged surface has mountainous features, including a globe-spanning mid-ocean ridge system, as well as undersea volcanoes, oceanic trenches, submarine canyons, oceanic plateaus and abyssal plains.

The mass of the oceans is approximately 1.35×10^{18} metric tons, or about 1/4400 of the total mass of the Earth. The oceans cover an area of 3.618×10^{8} km² with a mean depth of 3,682 m, resulting in an estimated volume of 1.332×10^{9} km³.

Biomass

One measure of the relative importance of different marine habitats is the rate at which they produce biomass.

Producer	Biomass productivity (gC/m²/yr)	Total area (million km²)	Total production (billion tonnes C/yr)
swamps and marshes	2,500		
coral reefs	2,000	0.28	0.56
algal beds	2,000		
river estuaries	1,800		
open ocean	125	311	39

Coastal

Coastlines can be volatile habitats.

Marine coasts are dynamic environments which constantly change, like the ocean which partially shape them. The Earth's natural processes, including weather and sea level change, result in the erosion, accretion and resculpturing of coasts as well as the flooding and creation of continental shelves and drowned river valleys.

The main agents responsible for deposition and erosion along coastlines are waves, tides and currents. The formation of coasts also depends on the nature of the rocks they are made of – the harder the rocks the less likely they are to erode, so variations in rock hardness result in coastlines with different shapes.

Tides often determine the range over which sediment is deposited or eroded. Areas with high tidal ranges allow waves to reach farther up the shore, and areas with lower tidal ranges produce deposition at a smaller elevation interval. The tidal range is influenced by the size and shape of the coastline. Tides do not typically cause erosion by themselves; however, tidal bores can erode as the waves surge up river estuaries from the ocean.

Waves erode coastline as they break on shore releasing their energy; the larger the wave the more energy it releases and the more sediment it moves. Sediment deposited by waves comes from eroded cliff faces and is moved along the coastline by the waves. Sediment deposited by rivers is the dominant influence on the amount of sediment located on a coastline.

Shores that look permanent through the short perceptive of a human lifetime are in fact among the most temporary of all marine structures.

The sedimentologist Francis Shepard classified coasts as primary or secondary:

- Primary coasts are shaped by non-marine processes, by changes in the land form. If a coast is in much the same condition as it was when sea level was stabilised after the last ice age, it is called a primary coast. "Primary coasts are created by erosion (the wearing away of soil or rock), deposition (the buildup of sediment or sand) or tectonic activity (changes in the structure of the rock and soil because of earthquakes). Many of these coastlines were formed as the sea level rose during the last 18,000 years, submerging river and glacial valleys to form bays and fjords." An example of a primary coast is a river delta, which forms when a river deposits soil and other material as it enters the sea.
- Secondary coasts are produced by marine processes, such as the action of the sea or by creatures that live in it. Secondary coastlines include sea cliffs, barrier islands, mud flats, coral reefs, mangrove swamps and salt marshes.

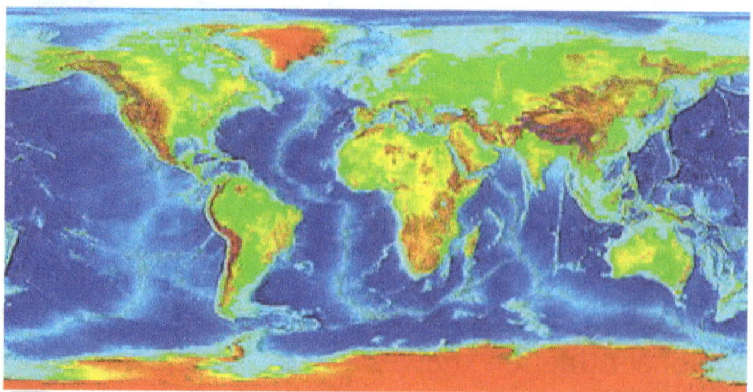

The global continental shelf, highlighted in cyan, defines the extent of coastal habitats, and occupies 5% of the total world area.

Continental coastlines usually have a continental shelf, a shelf of relatively shallow water, less than 200 metres deep, which extends 68 km on average beyond the coast.

Worldwide, continental shelves occupy a total area of about 24 million km² (9 million sq mi), 8% of the ocean's total area and nearly 5% of the world's total area. Since the continental shelf is usually less than 200 metres deep, it follows that coastal habitats are generally photic, situated in the sunlit epipelagic zone. This means the conditions for photosynthetic processes so important for primary production, are available to coastal marine habitats. Because land is nearby, there are large discharges of nutrient rich land runoff into coastal waters. Further, periodic upwellings from the deep ocean can provide cool and nutrient rich currents along the edge of the continental shelf.

As a result, coastal marine life is the most abundant in the world. It is found in tidal pools, fjords and estuaries, near sandy shores and rocky coastlines, around coral reefs and on or above the continental shelf. Coastal fish include small forage fish as well as the larger predator fish that feed on them. Forage fish thrive in inshore waters where high productivity results from upwelling and shoreline run off of nutrients. Some are partial residents that spawn in streams, estuaries and bays, but most complete their life cycle in the zone. There can also be a mutualism between species that occupy adjacent marine habitats. For example, fringing reefs just below low tide level have a mutually beneficial relationship with mangrove forests at high tide level and sea grass meadows in between: the reefs protect the mangroves and seagrass from strong currents and waves that would damage them or erode the sediments in which they are rooted, while the mangroves and seagrass protect the coral from large influxes of silt, fresh water and pollutants. This additional level of variety in the environment is beneficial to many types of coral reef animals, which for example may feed in the sea grass and use the reefs for protection or breeding.

Coastal habitats are the most visible marine habitats, but they are not the only important marine habitats. Coastlines run for 380,000 kilometres, and the total volume of the ocean is 1,370 million cu km. This means that for each metre of coast, there is 3.6 cu km of ocean space available somewhere for marine habitats.

Intertidal

Intertidal zones, those areas close to shore, are constantly being exposed and covered by the ocean's tides. A huge array of life lives within this zone.

Waves and currents shape the intertidal shoreline, eroding the softer rocks and transporting and grading loose particles into shingles, sand or mud.

Shore habitats range from the upper intertidal zones to the area where land vegetation takes prominence. It can be underwater anywhere from daily to very infrequently. Many species here are scavengers, living off of sea life that is washed up on the shore. Many land animals also make much use of the shore and intertidal habitats. A subgroup of organisms in this habitat bores and grinds exposed rock through the process of bioerosion.

Sandy Shores

Sandy shores provide shifting homes to many species.

Sandy shores, also called beaches, are coastal shorelines where sand accumulates. Waves and currents shift the sand, continually building and eroding the shoreline. Longshore currents flow parallel to the beaches, making waves break obliquely on the sand. These currents transport large amounts of sand along coasts, forming spits, barrier islands and tombolos. Longshore currents also commonly create offshore bars, which give beaches some stability by reducing erosion.

Sandy shores are full of life, The grains of sand host diatoms, bacteria and other microscopic creatures. Some fish and turtles return to certain beaches and spawn eggs in the sand. Birds habitat beaches, like gulls, loons, sandpipers, terns and pelicans. Aquatic mammals, such sea lions, recuperate on them. Clams, periwinkles, crabs, shrimp, starfish and sea urchins are found on most beaches.

Sand is a sediment made from small grains or particles with diameters between about 60 μm and 2 mm. Mud is a sediment made from particles finer than sand. This small particle size means that mud particles tend to stick together, whereas sand particles do not. Mud is not easily shifted by waves and currents, and when it dries out, cakes into a solid. By contrast, sand is easily shifted by waves and currents, and when sand dries out it can be blown in the wind, accumulating into shifting sand dunes. Beyond the high tide mark, if the beach is low-lying, the wind can form rolling hills of sand dunes. Small dunes shift and reshape under the influence of the wind while larger dunes stabilise the sand with vegetation.

Ocean processes grade loose sediments to particle sizes other than sand, such as gravel or cobbles. Waves breaking on a beach can leave a berm, which is a raised ridge of coarser pebbles or sand, at the high tide mark. Shingle beaches are made of particles larger than sand, such as cobbles, or small stones. These beaches make poor habitats. Little life survives because the stones are churned and pounded together by waves and currents.

Rocky Shores

Tidepools on rocky shores make turbulent habitats for many forms of marine life.

The relative solidity of rocky shores seems to give them a permanence compared to the shifting nature of sandy shores. This apparent stability is not real over even quite short geological time scales, but it is real enough over the short life of an organism. In contrast to sandy shores, plants and animals can anchor themselves to the rocks.

Competition can develop for the rocky spaces. For example, barnacles can compete successfully on open intertidal rock faces to the point where the rock surface is covered with them. Barnacles resist desiccation and grip well to exposed rock faces. However, in the crevices of the same rocks, the inhabitants are different. Here mussels can be the successful species, secured to the rock with their byssal threads.

Rocky and sandy coasts are vulnerable because humans find them attractive and want to live near them. An increasing proportion of the humans live by the coast, putting pressure on coastal habitats.

Mudflats

Mudflats become temporary habitats for migrating birds.

Mudflats are coastal wetlands that form when mud is deposited by tides or rivers. They are found in sheltered areas such as bays, bayous, lagoons, and estuaries. Mudflats may be viewed geologically as exposed layers of bay mud, resulting from deposition of estuarine silts, clays and marine animal detritus. Most of the sediment within a mudflat is within the intertidal zone, and thus the flat is submerged and exposed approximately twice daily.

Mudflats are typically important regions for wildlife, supporting a large population, although levels of biodiversity are not particularly high. They are of particular importance to migratory birds. In the United Kingdom mudflats have been classified as a Biodiversity Action Plan priority habitat.

Mangrove and Salt Marshes

Mangroves provide nurseries for fish.

Mangrove swamps and salt marshes form important coastal habitats in tropical and temperate areas respectively.

Mangroves are species of shrubs and medium size trees that grow in saline coastal sediment habitats in the tropics and subtropics – mainly between latitudes 25 °N and 25 °S. The saline conditions tolerated by various species range from brackish water, through pure seawater (30 to 40 ppt), to water concentrated by evaporation to over twice the salinity of ocean seawater (up to 90 ppt). There are many mangrove species, not all closely related. The term "mangrove" is used generally to cover all of these species, and it can be used narrowly to cover just mangrove trees of the genus *Rhizophora*.

Mangroves form a distinct characteristic saline woodland or shrubland habitat, called a *mangrove swamp* or *mangrove forest*. Mangrove swamps are found in depositional coastal environments, where fine sediments (often with high organic content) collect in areas protected from high-energy wave action. Mangroves dominate three quarters of tropical coastlines.

Estuaries

An estuary is a partly enclosed coastal body of water with one or more rivers or streams flowing into it, and with a free connection to the open sea. Estuaries form a transition zone between river environments and ocean environments and are subject to both marine influences, such as tides, waves, and the influx of saline water; and riverine influences, such as flows of fresh water and sediment. The inflow of both seawater and freshwater provide high levels of nutrients in both the water column and sediment, making estuaries among the most productive natural habitats in the world.

Estuaries occur when rivers flow into a coastal bay or inlet. They are nutrient rich and have a transition zone which moves from freshwater to saltwater.

Most estuaries were formed by the flooding of river-eroded or glacially scoured valleys when sea level began to rise about 10,000-12,000 years ago. They are amongst the most heavily populated areas throughout the world, with about 60% of the world's population living along estuaries and the coast. As a result, estuaries are suffering degradation by many factors, including sedimentation from soil erosion from deforestation; overgrazing and other poor farming practices; overfishing; drainage and filling of wetlands; eutrophication due to excessive nutrients from sewage and animal wastes; pollutants including heavy metals, PCBs, radionuclides and hydrocarbons from sewage inputs; and diking or damming for flood control or water diversion.

Estuaries provide habitats for a large number of organisms and support very high productivity. Estuaries provide habitats for salmon and sea trout nurseries, as well as migratory bird populations. Two of the main characteristics of estuarine life are the variability in salinity and sedimentation. Many species of fish and invertebrates have various methods to control or conform to the shifts in salt concentrations and are termed osmoconformers and osmoregulators. Many animals also burrow to avoid predation and to live in the more stable sedimental environment. However, large numbers of bacteria are found within the sediment which have a very high oxygen demand. This reduces the levels of oxygen within the sediment often resulting in partially anoxic conditions, which can be further exacerbated by limited water flux. Phytoplankton are key primary producers in estuaries. They move with the water bodies and can be flushed in and out with the tides. Their productivity is largely dependent on the turbidity of the water. The main phytoplankton present are diatoms and dinoflagellates which are abundant in the sediment.

Kelp Forests

Kelp forests provide habitat for many marine organisms.

Kelp forests are underwater areas with a high density of kelp. They form some of the most productive and dynamic ecosystems on Earth. Smaller areas of anchored kelp are called *kelp beds*. Kelp forests occur worldwide throughout temperate and polar coastal oceans.

Kelp forests provide a unique three-dimensional habitat for marine organisms and are a source for understanding many ecological processes. Over the last century, they have been the focus of extensive research, particularly in trophic ecology, and continue to provoke important ideas that are relevant beyond this unique ecosystem. For example, kelp forests can influence coastal oceanographic patterns and provide many ecosystem services.

However, humans have contributed to kelp forest degradation. Of particular concern are the effects of overfishing nearshore ecosystems, which can release herbivores from their normal population regulation and result in the over-grazing of kelp and other algae. This can rapidly result in transitions to barren landscapes where relatively few species persist.

Frequently considered an ecosystem engineer, kelp provides a physical substrate and habitat for kelp forest communities. In algae (Kingdom: Protista), the body of an individual organism is known as a thallus rather than as a plant (Kingdom: Plantae). The morphological structure of a kelp thallus is defined by three basic structural units:

- The holdfast is a root-like mass that anchors the thallus to the sea floor, though unlike true roots it is not responsible for absorbing and delivering nutrients to the rest of the thallus;
- The stipe is analogous to a plant stalk, extending vertically from the holdfast and providing a support framework for other morphological features;

- The fronds are leaf- or blade-like attachments extending from the stipe, sometimes along its full length, and are the sites of nutrient uptake and photosynthetic activity.

In addition, many kelp species have pneumatocysts, or gas-filled bladders, usually located at the base of fronds near the stipe. These structures provide the necessary buoyancy for kelp to maintain an upright position in the water column.

The environmental factors necessary for kelp to survive include hard substrate (usually rock), high nutrients (e.g., nitrogen, phosphorus), and light (minimum annual irradiance dose > 50 E m^{-2}). Especially productive kelp forests tend to be associated with areas of significant oceanographic upwelling, a process that delivers cool nutrient-rich water from depth to the ocean's mixed surface layer. Water flow and turbulence facilitate nutrient assimilation across kelp fronds throughout the water column. Water clarity affects the depth to which sufficient light can be transmitted. In ideal conditions, giant kelp (*Macrocystis spp.*) can grow as much as 30-60 centimetres vertically per day. Some species such as *Nereocystis* are annual while others like *Eisenia* are perennial, living for more than 20 years. In perennial kelp forests, maximum growth rates occur during upwelling months (typically spring and summer) and die-backs correspond to reduced nutrient availability, shorter photoperiods and increased storm frequency.

Seagrass Meadows

White-spotted puffers like living in seagrass areas.

Seagrasses are flowering plants from one of four plant families which grow in marine environments. They are called *seagrasses* because the leaves are long and narrow and are very often green, and because the plants often grow in large meadows which look like grassland. Since seagrasses photosynthesize and are submerged, they must grow submerged in the photic zone, where there is enough sunlight. For this reason, most occur in shallow and sheltered coastal waters anchored in sand or mud bottoms.

Seagrasses form extensive beds or meadows, which can be either monospecific (made up of one species) or multispecific (where more than one species co-exist). Seagrass beds make highly diverse and productive ecosystems. They are home to phyla such as juvenile and adult fish, epiphytic and free-living macroalgae and microalgae, mollusks, bristle worms, and nematodes. Few species were originally considered to feed directly on seagrass leaves (partly because of their low nutritional content), but scientific reviews and improved working methods have shown that seagrass herbivory is a highly important link in the food chain, with hundreds of species feeding on seagrasses worldwide, including green turtles, dugongs, manatees, fish, geese, swans, sea urchins and crabs.

Seagrasses are ecosystem engineers in the sense that they partly create their own habitat. The leaves slow down water-currents increasing sedimentation, and the seagrass roots and rhizomes stabilize the seabed. Their importance to associated species is mainly due to provision of shelter (through their three-dimensional structure in the water column), and due to their extraordinarily high rate of primary production. As a result, seagrasses provide coastal zones with ecosystem services, such as fishing grounds, wave protection, oxygen production and protection against coastal erosion. Seagrass meadows account for 15% of the ocean's total carbon storage.

Coral Reefs

Reefs comprise some of the densest and most diverse habitats in the world. The best-known types of reefs are tropical coral reefs which exist in most tropical waters; however, reefs can also exist in cold water. Reefs are built up by corals and other calcium-depositing animals, usually on top of a rocky outcrop on the ocean floor. Reefs can also grow on other surfaces, which has made it possible to create artificial reefs. Coral reefs also support a huge community of life, including the corals themselves, their symbiotic zooxanthellae, tropical fish and many other organisms.

Much attention in marine biology is focused on coral reefs and the El Niño weather phenomenon. In 1998, coral reefs experienced the most severe mass bleaching events on record, when vast expanses of reefs across the world died because sea surface temperatures rose well above normal. Some reefs are recovering, but scientists say that between 50% and 70% of the world's coral reefs are now endangered and predict that global warming could exacerbate this trend.

Open Ocean

The open ocean is relatively unproductive because of a lack of nutrients, yet because it is so vast, it has more overall primary production than any other marine habitat. Only about 10 percent of marine species live in the open ocean. But among them are the largest and fastest of all marine animals, as well as the animals that dive the deepest and migrate the longest. In the depths lurk animal that, to our eyes, appear hugely alien.

Surface Waters

In the open ocean, sunlit surface epipelagic waters get enough light for photosynthesis, but there are often not enough nutrients. As a result, large areas contain little life apart from migrating animals.

The surface waters are sunlit. The waters down to about 200 metres are said to be in the epipelagic zone. Enough sunlight enters the epipelagic zone to allow photosynthesis by phytoplankton. The epipelagic zone is usually low in nutrients. This partially because the organic debris produced in the zone, such as excrement and dead animals, sink to the depths and are lost to the upper zone. Photosynthesis can happen only if both sunlight and nutrients are present.

In some places, like at the edge of continental shelves, nutrients can upwell from the ocean depth, or land runoff can be distributed by storms and ocean currents. In these areas, given that both sunlight and nutrients are now present, phytoplankton can rapidly establish itself, multiplying so fast that the water turns green from the chlorophyll, resulting in an algal bloom. These nutrient rich surface waters are among the most biologically productive in the world, supporting billions of tonnes of biomass.

Phytoplankton are eaten by zooplankton - small animals which, like phytoplankton, drift in the ocean currents. The most abundant zooplankton species are copepods and krill: tiny crustaceans that are the most numerous animals on Earth. Other types of zooplankton include jelly fish and the larvae of fish, marine worms, starfish, and other marine organisms". In turn, the zooplankton are eaten by filter-feeding animals, including some seabirds, small forage fish like herrings and sardines, whale sharks, manta rays, and the largest animal in the world, the blue whale. Yet again, moving up the foodchain, the small forage fish are in turn eaten by larger predators, such as tuna, marlin, sharks, large squid, seabirds, dolphins, and toothed whales.

Deep Sea

The layers of the Pelagic Zone.

The deep sea starts at the aphotic zone, the point where sunlight loses most of its energy in the water. Many life forms that live at these depths have the ability to create their own light a unique evolution known as bio-luminescence.

In the deep ocean, the waters extend far below the epipelagic zone, and support very different types of pelagic life forms adapted to living in these deeper zones.

Much of the aphotic zone's energy is supplied by the open ocean in the form of detritus. In deep water, marine snow is a continuous shower of mostly organic detritus falling from the upper layers of the water column. Its origin lies in activities within the productive photic zone. Marine snow includes dead or dying plankton, protists (diatoms), fecal matter, sand, soot and other inorganic dust. The "snowflakes" grow over time and may reach several centimetres in diameter, travelling for weeks before reaching the ocean floor. However, most organic components of marine snow are consumed by microbes, zooplankton and other filter-feeding animals within the first 1,000 metres of their journey, that is, within the epipelagic zone. In this way marine snow may be considered the foundation of deep-sea mesopelagic and benthic ecosystems: As sunlight cannot reach them, deep-sea organisms rely heavily on marine snow as an energy source.

Some deep-sea pelagic groups, such as the lanternfish, ridgehead, marine hatchetfish, and lightfish families are sometimes termed *pseudoceanic* because, rather than having an even distribution in open water, they occur in significantly higher abundances around structural oases, notably seamounts and over continental slopes. The phenomenon is explained by the likewise abundance of prey species which are also attracted to the structures.

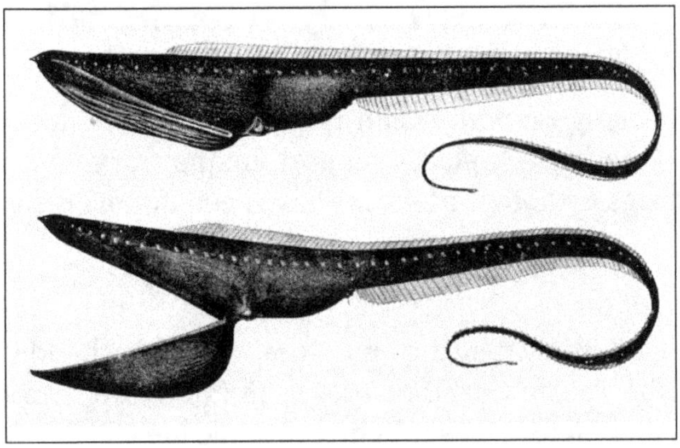

The umbrella mouth gulper eel can swallow a fish much larger than itself.

The fish in the different pelagic and deep water benthic zones are physically structured, and behave in ways, that differ markedly from each other. Groups of coexisting species within each zone all seem to operate in similar ways, such as the small mesopelagic vertically migrating plankton-feeders, the bathypelagic anglerfishes, and the deep water benthic rattails.

Ray finned species, with spiny fins, are rare among deep sea fishes, which suggests that deep sea fish are ancient and so well adapted to their environment that invasions by more modern fishes have been unsuccessful. The few ray fins that do exist are mainly in the Beryciformes and Lampriformes, which are also ancient forms. Most deep sea pelagic fishes belong to their own orders, suggesting a long evolution in deep sea environments. In contrast, deep water benthic species, are in orders that include many related shallow water fishes.

The umbrella mouth gulper is a deep sea eel with an enormous loosely hinged mouth. It can open its mouth wide enough to swallow a fish much larger than itself, and then expand its stomach to accommodate its catch.

Sea Floor

Vents and Seeps

Zooarium chimney provides a habitat for vent biota.

Hydrothermal vents along the mid-ocean ridge spreading centers act as oases, as do their opposites, cold seeps. Such places support unique marine biomes and many new marine microorganisms and other lifeforms have been discovered at these locations.

Trenches

The deepest recorded oceanic trenches measure to date is the Mariana Trench, near the Philippines, in the Pacific Ocean at 10,924 m (35,838 ft). At such depths, water pressure is extreme and there is no sunlight, but some life still exists. A white flatfish, a shrimp and a jellyfish were seen by the American crew of the bathyscaphe *Trieste* when it dove to the bottom in 1960.

Seamounts

Marine life also flourishes around seamounts that rise from the depths, where fish and other sea life congregate to spawn and feed.

ANTHROPOGENIC BIOMES

Anthropogenic biomes, also known as anthromes or human biomes are the global ecological patterns created by humans. Anthromes characterize the globally significant ecological patterns created by human populations and their use of land over the long term. Like the classic biomes that differentiate tropical rainforests, grasslands, tundra and other climate-driven ecological patterns across the terrestrial biosphere, anthropogenic biomes differentiate the human-driven ecological patterns across the terrestrial biosphere, including urban areas, villages and croplands.

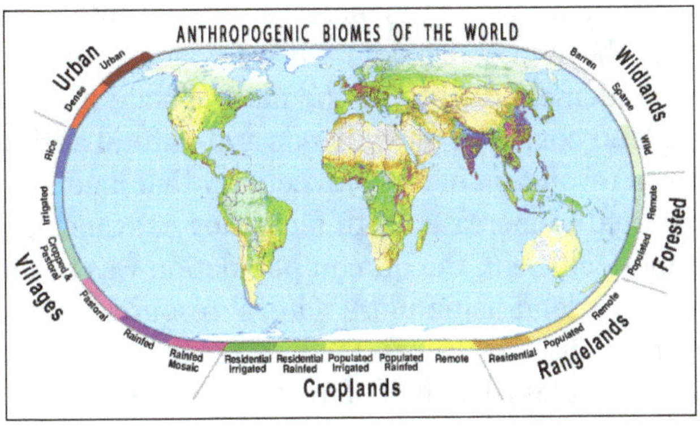

Anthromes offer a new way forward for ecology and earth science by recognizing and characterizing the vast global extent and ecological richness of human interactions with ecosystems across the terrestrial biosphere, going beyond a single dimension of human impact or disturbance.

Anthropogenic Transformation of the Biosphere

For more than a century, the biosphere has been described in terms of global ecosystem units called biomes, which are vegetation types like tropical rainforests and grasslands that are identified in relation to global climate patterns. Considering that human populations and their use of land have fundamentally altered global patterns of ecosystem form, process, and biodiversity, anthropogenic biomes provide a framework for integrating human systems with the biosphere in the Anthropocene.

Before 1700

Humans have been altering ecosystems since we have evolved. Evidence suggests that our ancestors were burning land to clear it at one million years ago. 600,000 years ago, humans were using spears to kill horses and other large animals in Great Britain and China. For the past tens of thousands of years, humans have greatly changed the plant and animal life around the globe, from what type of wildlife and plant life dominated to what type of ecosystems dominate. Examples include Native Americans; they altered the forest, burnt land to clear it, settled in cities, disrupting forests and other ecosystems, and built monuments that required moving large amounts of earth, such as the Cahokia Monuments. More examples are the civilizations of the ancient world; they mined large amounts of material, made roads, and especially for the Romans, when mining lead, released large amounts of Mercury and lead into the air. They also moved large numbers of animals for entertainment, disrupting wildlife patterns.

Agriculture (1700–present)

Humans have been altering ecosystems since before agriculture first developed, and as the human population has grown and become more technologically advanced over time, the land use for agricultural purposes has increased significantly. The anthropogenic biome in the 1700s, before the industrial revolution, was made up of mostly wild, untouched land, with no human settlement disturbing the natural state. In this time period, most of the Earth's ice-free land consisted of wildlands and natural anthromes, and it wasn't until after the industrial revolution in the 19th century that land use for agriculture and human settlements started to increase. With technology advancing and manufacturing processes becoming more efficient, the human population was beginning to thrive, and was subsequently requiring and using more natural resources. By the year 2000, over half of the Earth's ice free land was transformed into rangelands, croplands, villages and dense settlements, which left less than half of the Earth's land untouched. Anthropogenic changes between 1700 and 1800 were far smaller than those of the following centuries, and as such the rate of change has increased over time. As a result, the 20th century had the fastest rate of anthropogenic ecosystem transformation of the past 300 years.

Land Distribution

As the human population steadily increased in numbers throughout history, the use of natural resources and land began to increase, and the distribution of land used for various agricultural and settlement purposes began to change. The use of land around the world was transformed from its natural state to land used for agriculture, settlements and pastures to sustain the population and its growing needs. The distribution of land among anthromes underwent a shift away from natural anthromes and wildlands towards human-altered anthromes we are familiar with today. Now, the most populated anthromes (dense settlements and villages) account for only a small fraction of the global ice-free land. From the year 1700-2000, lands used for agriculture and urban settlements increased significantly, however the area occupied by rangelands increased even more rapidly, so that it became the dominant anthrome in the 20th century. As a result, the biggest global land-use change as a result of the industrial revolution, was the expansion of pastures.

Human Population

Following the industrial revolution, the human population experienced a rapid increase. The human population density in certain anthromes began to change, shifting away from rural environments to urban settlements, where the population density was much higher. These changes in population density between areas shifted global patterns of anthrome emergence, and also had wide-spread effects on various ecosystems. Half of the Earth's population now lives in cities, and most people reside in urban anthromes, with some populations dwelling in smaller cities and towns. Currently, human populations are expected to grow until at least midcentury, and the transformation of the Earth's anthromes are expected to follow this growth.

Current State of the Anthropogenic Biosphere

The present state of the terrestrial biosphere is predominantly anthropogenic. More than half of the terrestrial biosphere remains unused directly for agriculture or urban settlements, and of these unused lands still remaining, less than half are wildlands. Most of Earth's unused lands are now within the agricultural and settled landscapes of semi-natural, rangeland, cropland and village anthromes.

Major Anthromes

Anthromes include dense settlements (urban and mixed settlements), villages, croplands, rangelands and semi-natural lands and have been mapped globally using two different classification systems, There are currently 18 anthropogenic biomes.

Dense Settlements

Dense Settlements are the second most densely populated regions in the world. They are defined as areas with a high population density, though the density can be variable. The Population density, however, never falls below 100 persons/km, even in the non-urban parts of the dense settlements, and it has been suggested that these areas consist of both the edges of major cities in underdeveloped nations, and the long standing small towns throughout western Europe and Asia. Most often we think of dense settlements as cities, but dense settlements can also be suburbs, towns and rural settlements with high but fragmented populations.

Croplands

Croplands are another major anthrome throughout the world. Croplands include most of the cultivated lands of the world, and also about a quarter of global tree cover. Croplands which are locally irrigated have the highest human population density, likely due to the fact that it provides crops with a constant supply on water. This makes harvest time and crop survival more predictable. Croplands that are sustained mainly from the local rainfall are the most extensive of the populated anthromes, with annual precipitation near 1000 mm in certain areas of the globe. In these areas, there is sufficient water supplied by the climate to support all aspects of life without hardly any irrigation. However, in dryer areas, this method of agriculture would not be as productive.

Rangelands

Rangelands are a very broad anthropogenic biome group that has been described according to three levels of population density: residential, populated and remote. The Residential rangeland anthrome has two key features: its population density is never below 10 persons per square kilometre, and a substantial portion of its area is used for pasture. Pastures in rangelands are the most dominant land cover. Bare earth is significant in this anthrome, covering nearly one third of the land for every one square kilometer. Rangeland anthromes are less altered than croplands, but their alteration tends to increase with population. Domesticated grazing livestock are typically adapted to grasslands and savannas, so the alteration of these biomes tends to be less noticeable.

Forests

Forested anthromes are dominated by tree cover, and they have high precipitation and minimal human populations, where the population density is usually less than 3 persons/km^2. Most populated forests act as carbon sinks because of the lack of human activity. Without harmful emissions being released in the forests due to human activity, the vegetation is able to utilize carbon dioxide in the atmosphere, and act as a sink. Remote forests are a little different than populated forests because the majority of the vegetation in these forests have been clear-cut for human consumption. Forests are generally cleared to sustain substantial populations of domestic livestock, and to utilize the lumber.

Indoor

Very few biologists have studied the evolutionary processes at work in indoor environments. Estimates of the extent of residential and commercial buildings range between 1.3% and 6% of global ice-free land area. This area is just as extensive as other small biomes such as flooded grass-lands and tropical coniferous forests. The indoor biome is rapidly expanding, while forest anthromes are shrinking. The indoor biome of Manhattan is almost three times as large, in terms of its floor space, as is the geographical area of the island itself, due to the buildings rising up instead of spreading out. Thousands of species live in the indoor biome, many of them preferentially or even obligatorily. The only action that humans take to alter the evolution of the indoor biome is with cleaning practices. The field of indoor biomes will continue to change as long as our culture will change.

Aquatic

Managed aquatic biomes or aquatic anthromes have rarely been studied as such. They range from fish ponds, marine shrimp and bentic farming sites to large tracts of land such as parts of the Guadalquivir Marshes in Andalusia, Spain.

Implications of an Anthropogenic Biosphere

Humans have fundamentally altered global patterns of biodiversity and ecosystem processes. It is no longer possible to explain or predict ecological patterns or processes across the Earth without considering the human role. Human societies began transforming terrestrial ecology more than 50000 years ago, and evolutionary evidence has been presented demonstrating that the ultimate causes of human transformation of the biosphere are social and cultural, not biological, chemical, or physical. Anthropogenic biomes offer a new way forward by acknowledging human influence on global ecosystems and moving us toward models and investigations of the terrestrial biosphere that integrate human and ecological systems.

Challenges Facing Biodiversity in the Anthropogenic Biosphere

Extinctions

Over the past century, anthrome extent and land use intensity increased rapidly together with growing human populations, leaving wildlands without human population or land use in less than one quarter of the terrestrial biosphere. This massive transformation of Earth's ecosystems for human use has occurred with enhanced rates of species extinctions. Humans are directly causing species extinctions, especially of megafauna, by reducing, fragmenting and transforming native habitats and by overexploiting individual species. Current rates of extinctions vary greatly by taxa, with mammals, reptiles and amphibians especially threatened; however there is growing evidence that viable populations of many, if not most native taxa, especially plants, may be sustainable within anthromes. With the exception of especially vulnerable taxa, the majority of native species may be capable of maintaining viable populations in anthromes.

Conservation

Anthromes present an alternative view of the terrestrial biosphere by characterizing the diversity of global ecological land cover patterns created and sustained by human population densities and land use while also incorporating their relationships with biotic communities. Biomes and ecoregions are limited in that they reduce human influences, and an increasing number of conservation biologists have argued that biodiversity conservation must be extended to habitats directly shaped by humans. Within anthromes, including densely populated anthromes, humans rarely use all available land. As a result, anthromes are generally mosaics of heavily used lands and less intensively used lands. Protected areas and biodiversity hotspots are not distributed equally across anthromes. Less populated anthromes contain a greater proportion of protected areas. While 23.4% of remote woodland anthrome is protected, only 2.3% of irrigated village anthrome is protected. There is increasing evidence that suggests that biodiversity conservation can be effective in both densely and sparsely settled anthromes. Land sharing and land sparing are increasingly seen as conservation strategies.

MICROBIAL BIOMES

A microbiota is an "ecological community of commensal, symbiotic and pathogenic microorganisms" found in and on all multicellular organisms studied to date from plants to animals. A microbiota includes bacteria, archaea, protists, fungi and viruses. Microbiota have been found to be crucial for immunologic, hormonal and metabolic homeostasis of their host. The synonymous term microbiome describes either the collective genomes of the microorganisms that reside in an environmental niche or the microorganisms themselves.

The microbiome and host emerged during evolution as a synergistic unit from epigenetics and genetic characteristics, sometimes collectively referred to as a holobiont.

All plants and animals, from simple life forms to humans, live in close association with microbial organisms. Several advances have driven the perception of microbiomes, including:

- The ability to perform genomic and gene expression analyses of single cells and of entire microbial communities in the disciplines of metagenomics and metatranscriptomics.
- Databases accessible to researchers across multiple disciplines.
- Methods of mathematical analysis suitable for complex data sets.

Biologists have come to appreciate that microbes make up an important part of an organism's phenotype, far beyond the occasional symbiotic case study.

Types of Host Relationships

Commensalism, a concept developed by Pierre-Joseph van Beneden, a Belgian professor at the University of Louvain during the nineteenth century is central to the microbiome, where microbiota colonize a host in a non-harmful coexistence. The relationship with their host is called mutualistic when organisms perform tasks that are known to be useful for the host, parasitic, when disadvantageous to the host. Other authors define a situation as mutualistic where both benefit, and commensal, where the unaffected host benefits the symbiont. A nutrient exchange may be bidirectional or unidirectional, may be context dependent and may occur in diverse ways. Microbiota that are expected to be present, and that under normal circumstances do not cause disease, are deemed normal flora or normal microbiota.

Acquisition and Change

The initial acquisition of microbiota in animals from mammalians to marine sponges is at birth, and may even occur through the germ cell line. In plants, the colonizing process can be initiated below ground in the root zone, around the germinating seed, the spermosphere, or originate from the above ground parts, the phyllosphere and the flower zone or anthosphere. The stability of the rhizosphere microbiota over generations depends upon the plant type but even more on the soil composition, i.e. living and non living environment.

Microbiota by Host

Consensus exists among evolutionary biologists that one should not separate an organism's genes from the context of its resident microbes.

Humans

The human microbiota includes bacteria, fungi, archaea and viruses. Micro-animals which live on the human body are excluded. The human microbiome refers to their genomes.

Humans are colonized by many microorganisms; the traditional estimate was that humans live with ten times more non-human cells than human cells; more recent estimates have lowered this to 3:1 and even to about 1:1.

The Human Microbiome Project sequenced the genome of the human microbiota, focusing particularly on the microbiota that normally inhabit the skin, mouth, nose, digestive tract, and vagina. It reached a milestone in 2012 when it published initial results.

Non-human Animals

Amphibians have microbiota on their skin. Some species are able to carry a fungus named *Batrachochytrium dendrobatidis*, which in others can cause a deadly infection Chytridiomycosis depending on their microbiome, resisting pathogen colonization or inhibiting their growth with antimicrobial skin peptides.

In mammals, herbivores such as cattle depend on their rumen microbiome to convert cellulose into proteins, short chain fatty acids, and gases. Culture methods cannot provide information on all microorganisms present. Comparative metagenomic studies yielded the surprising result that individual cattle possess markedly different community structures, predicted phenotype, and metabolic potentials, even though they were fed identical diets, were housed together, and were apparently functionally identical in their utilization of plant cell wall resources.

Mice have become the most studied mammalian regarding their microbiomes. The gut microbiota have been studied in relation to allergic airway disease, obesity, gastrointestinal diseases and diabetes. Perinatal shifting of microbiota through low dose antibiotics can have long-lasting effects on future susceptibility to allergic airway disease.

The frequency of certain subsets of microbes has been linked to disease severity. The presence of specific microbes early in postnatal life, instruct future immune responses. In gnotobiotic mice certain gut bacteria were found to transmit a particular phenotype to recipient germ-free mice, that promoted accumulation of colonic regulatory T cells, and strains that modulated mouse adiposity and cecal metabolite concentrations. This combinatorial approach enables a systems-level understanding of microbial contributions to human biology. But also other mucoide tissues as lung and vagina have been studied in relation to diseases such as asthma, allergy and vaginosis.

Insects have their own microbiomes. For example, leaf-cutter ants form huge underground colonies harvesting hundreds of kilograms of leaves each year and are unable to digest the cellulose in the leaves directly. They maintain fungus gardens as the colony's primary food source. While the fungus itself does not digest cellulose, a microbial community containing a diversity of bacteria is doing so. Analysis of the microbial population's genome revealed many genes with a role in cellulose digestion. This microbiome's predicted carbohydrate-degrading enzyme profile is similar to that of the bovine rumen, but the species composition is almost entirely different. Gut microbiota of the fruit fly can affect the way its gut looks, by impacting epithelial renewal rate, cellular spacing, and the composition of different cell types in the epithelium. When the moth Spodoptera exigua is infected with baculovirus immune-related genes are downregulated and the amount of its gut microbiota increases.

Plants

Plants are attractive hosts for microorganisms since they provide a variety of nutrients. Microorganisms on plants can be epiphytes (found on the plants) or endophytes (found inside plant tissue). Oomycetes and fungi have, through convergent evolution, developed similar morphology and occupy similar ecological niches. They develop hyphae, threadlike structures that penetrate the host cell. In mutualistic situations the plant often exchanges hexose sugars for inorganic phosphate from the fungal symbiont. It is speculated that such very ancient associations have aided plants when they first colonized land. Plant-growth promoting bacteria (PGPB) provide the plant with essential services such as nitrogen fixation, solubilization of minerals such as phosphorus, synthesis of plant hormones, direct enhancement of mineral uptake, and protection from pathogens. PGPBs may protect plants from pathogens by competing with the pathogen for an ecological niche or a substrate, producing inhibitory allelochemicals, or inducing systemic resistance in host plants to the pathogen.

Research

The symbiotic relationship between a host and its microbiota is under laboratory research for how it may shape the immune system of mammals. In many animals, the immune system and microbiota may engage in "cross-talk" by exchanging chemical signals, which may enable the microbiota to influence immune reactivity and targeting.

Bacteria can be transferred from mother to child through direct contact and after birth. As the infant microbiome is established, commensal bacteria quickly populate the gut, prompting a range of immune responses and "programming" the immune system with long-lasting effects. The bacteria are able to stimulate lymphoid tissue associated with the gut mucosa, which enables the tissue to produce antibodies for pathogens that may enter the gut.

The human microbiome may play a role in the activation of toll-like receptors in the intestines, a type of pattern recognition receptor host cells use to recognize dangers and repair damage. Pathogens can influence this coexistence leading to immune dysregulation including and susceptibility to diseases, mechanisms of inflammation, immune tolerance, and autoimmune diseases.

Co-evolution of Microbiota

Organisms evolve within eco-systems so that the change of one organism affects the change of others. The Hologenome theory of evolution proposes that an object of natural selection is not the individual organism, but the organism together with its associated organisms, including its microbial communities.

Coral reefs: The hologenome theory originated in studies on coral reefs. Coral reefs are the largest structures created by living organisms, and contain abundant and highly complex microbial communities. Over the past several decades, major declines in coral populations have occurred. Climate change, water pollution and over-fishing are three stress factors that have been described as leading to disease susceptibility. Over twenty different coral diseases have been described, but of these, only a handful have had their causative agents isolated and characterized. Coral bleaching is the most serious of these diseases. In the Mediterranean Sea, the bleaching of *Oculina patagonica* was first described in 1994 and shortly determined to be due to infection by Vibrio shiloi. From 1994 to 2002, bacterial bleaching of *O. patagonica* occurred every summer in the eastern Mediterranean. Surprisingly, however, after 2003, *O. patagonica* in the eastern Mediterranean has been resistant to *V. shiloi* infection, although other diseases still cause bleaching. The surprise stems from the knowledge that corals are long lived, with lifespans on the order of decades, and do not have adaptive immune systems. Their innate immune systems do not produce antibodies, and they should seemingly not be able to respond to new challenges except over evolutionary time scales.

The puzzle of how corals managed to acquire resistance to a specific pathogen led to a 2007 proposal, that a dynamic relationship exists between corals and their symbiotic microbial communities. It is thought that by altering its composition, the holobiont can adapt to changing environmental conditions far more rapidly than by genetic mutation and selection alone. Extrapolating this hypothesis to other organisms, including higher plants and animals, led to the proposal of the "hologenome theory of evolution".

As of 2007 the hologenome theory was still being debated. A major criticism has been the claim that *V. shiloi* was misidentified as the causative agent of coral bleaching, and that its presence in bleached *O. patagonica* was simply that of opportunistic colonization. If this is true, the basic observation leading to the theory would be invalid. The theory has gained significant popularity as a way of explaining rapid changes in adaptation that cannot otherwise be explained by traditional mechanisms of natural selection. Within the hologenome theory, the holobiont has not only become the principal unit of natural selection but also the result of other step of integration that it is also observed at the cell (symbiogenesis, endosymbiosis) and genomic levels.

Research Methods

Targeted Amplicon Sequencing

Targeted amplicon sequencing relies on having some expectations about the composition of the community that is being studied. In target amplicon sequencing a phylogenetically informative marker is targeted for sequencing. Such a marker should be present in ideally all the expected organisms. It should also evolve in such a way that it is conserved enough that primers can target genes from a wide range of organisms while evolving quickly enough to allow for finer resolution at the taxonomic level. A common marker for human microbiome studies is the gene for bacterial 16S rRNA (*i.e.* "16S rDNA", the sequence of DNA which encodes the ribosomal RNA molecule). Since ribosomes are present in all living organisms, using 16S rDNA allows for DNA to be amplified from many more organisms than if another marker were used. The 16S rDNA gene contains both slowly evolving regions and fast evolving regions; the former can be used to design broad primers while the latter allow for finer taxonomic distinction. However, species-level resolution is not typically possible using the 16S rDNA. Primer selection is an important step, as anything that cannot be targeted by the primer will not be amplified and thus will not be detected. Different sets of primers have been shown to amplify different taxonomic groups due to sequence variation.

Targeted studies of eukaryotic and viral communities are limited and subject to the challenge of excluding host DNA from amplification and the reduced eukaryotic and viral biomass in the human microbiome.

After the amplicons are sequenced, molecular phylogenetic methods are used to infer the composition of the microbial community. This is done by clustering the amplicons into Operational Taxonomic Units (OTUs) and inferring phylogenetic relationships between the sequences. Due to the complexity of the data, distance measures such as UniFrac distances are usually defined between microbiome samples, and downstream multivariate methods are carried out on the distance matrices. An important point is that the scale of data is extensive, and further approaches must be taken to identify patterns from the available information. Tools used to analyze the data include VAMPS, QIIME and mothur.

Metagenomic Sequencing

Metagenomics is also used extensively for studying microbial communities. In metagenomic sequencing, DNA is recovered directly from environmental samples in an untargeted manner with the goal of obtaining an unbiased sample from all genes of all members of the community. Recent studies use shotgun Sanger sequencing or pyrosequencing to recover the sequences of the reads. The reads can then be assembled into contigs. To determine the phylogenetic identity of a sequence, it is compared to available full genome sequences using methods such as BLAST. One drawback of this approach is that many members of microbial communities do not have a representative sequenced genome, but this applies to 16S rRNA amplicon sequencing as well and is a fundamental problem. With shotgun sequencing, it can be resolved by having a high coverage (50-100x) of the unknown genome, effectively doing a de novo genome assembly. As soon as there is a complete genome of an unknown organism available it can be compared phylogenetically and the organism put into its place in the tree of life, by creating new taxa. An emerging approach is to combine shotgun sequencing with proximity-ligation data (Hi-C) to assemble complete microbial genomes without culturing.

Despite the fact that metagenomics is limited by the availability of reference sequences, one significant advantage of metagenomics over targeted amplicon sequencing is that metagenomics data can elucidate the functional potential of the community DNA. Targeted gene surveys cannot do this as they only reveal the phylogenetic relationship between the same gene from different organisms. Functional analysis is done by comparing the recovered sequences to databases of metagenomic annotations such as KEGG. The metabolic pathways that these genes are involved in can then be predicted with tools such as MG-RAST, CAMERA and IMG/M.

RNA and Protein-based Approaches

Metatranscriptomics studies have been performed to study the gene expression of microbial communities through methods such as the pyrosequencing of extracted RNA. Structure based studies have also identified non-coding RNAs (ncRNAs) such as ribozymes from microbiota. Metaproteomics is an approach that studies the proteins expressed by microbiota, giving insight into its functional potential.

ENDOLITHIC BIOMES

The endolithic (meaning “within rock”) biome is a biome located below the surface of the Earth. Other examples of biomes include savanna, desert, and jungle. The endolithic biome consists entirely of microbes living in tiny pores and cracks of rocks, and extends down to at least 3 km (9,600 feet) below the surface. The endolithic biome may be considered the most harsh biome on Earth, due to a near-complete absence of sunlight, oxygen, and most nutrients. Endolithic microbes have affectionately been referred to as "bugs from hell."

The endolithic biome is the most recently discovered biome. Serious exploration only began in the early 1990s. The most popular venue for study of these microbes are gold mines, some of which extend as far as two miles below the surface. Microbes found are either bacteria or archaea. All microbes in the endolithic biome are extremophiles, meaning they thrive in extreme conditions of heat or absence of nutrients.

Microbes in the endolithic biome reproduce very slowly, due to the scarce nutrients. Much energy is expended on repairing damage from cosmic rays, as well. Preliminary studies suggest that some species may engage in cell division as rarely as once every 100 years. By contrast, certain surface microbes replicate every 30 minutes.

Investigators of endolithic biomes report that the main limiting factor in microbe penetration underground is temperature rather than rock density or chemical factors. Judging from hyperthermophile ("super-heat-loving") organisms such as the recently discovered Strain 121, capable of reproducing at a temperature of 121 °C (250 °F), the temperature of an autoclave, endolithic organisms may exist as far as 4.5 km (2.8 mi) below the continental crust, and 7.5 km (4.6 mi) below the ocean floor. Due to the cost involved in digging so deep, no serious investigations of microbe life at such depths has yet been carried out. But scientists suspect that the endolithic biome may account for a substantial percentage of global biomass.

Because endolithic biome conditions resemble those that might be found on planets such as Mars or the moons of Jupiter, studies of the biome have been funded by agencies such as NASA, with an interest in their implications for the possibility of extraterrestrial life.

References

- What-are-biomes, biomes, science: softschools.com, Retrieved 30 August, 2019
- Sherwood, linda; willey, joanne; woolverton, christopher (2013). Prescott's microbiology (9th ed.). New york: mcgraw hill. Pp. 713–721. Isbn 9780073402406. Oclc 886600661
- "The vocabulary of microbiome research: a proposal". Microbiome. 3: 31. Doi:10.1186/s40168-015-0094-5. Pmc 4520061. Pmid 26229597
- The world's oceans and seas. Archived 2006-02-24 at the wayback machine encarta. Retrieved 19 april 2008
- Rona, peter a. (2003). "resources of the sea floor". Science. 299 (5607): 673–674. Doi:10.1126/science.1080679. Pmid 12560541. Retrieved 2007-02-04
- Ricklefs, robert e.; miller, gary leon (2000). Ecology (4th ed.). Macmillan. P. 192. Isbn 978-0-7167-2829-0

- Anthrome-anthropogenic-biome: blogspot.com, Retrieved 16 January, 2019
- Ellis, e. C.; ramankutty, n. (2008). "putting people in the map: anthropogenic biomes of the world". Frontiers in ecology and the environment. 6(8): 439–447. Doi:10.1890/070062
- Chapin iii, f. Stuart; matson, pamela a; vitousek, peter m (2012). Principles of terrestrial ecosystem ecology (second ed.). Springer. Isbn 978-1-4419-9504-9
- What-is-the-endolithic-biome: wisegeek.com, Retrieved 29 March, 2019
- Salvucci, e. (2016). "microbiome, holobiont and the net of life". Critical reviews in microbiology. 42 (3): 485–494. Doi:10.3109/1040841x.2014.962478. Pmid 25430522

4

Classification of Biosphere

Biosphere refers to the global ecosystem of living organisms and non-living factors. It can be classified into various zones such as longitudes and latitudes, Equator, deep biosphere, Arctic circles and Antarctic circles. The aim of this chapter is to explore the classifications of biosphere, which are crucial for understanding the subject in its entirety.

LATITUDE AND LONGITUDE

Latitude and longitude is a coordinate system by means of which the position or location of any place on Earth's surface can be determined and described.

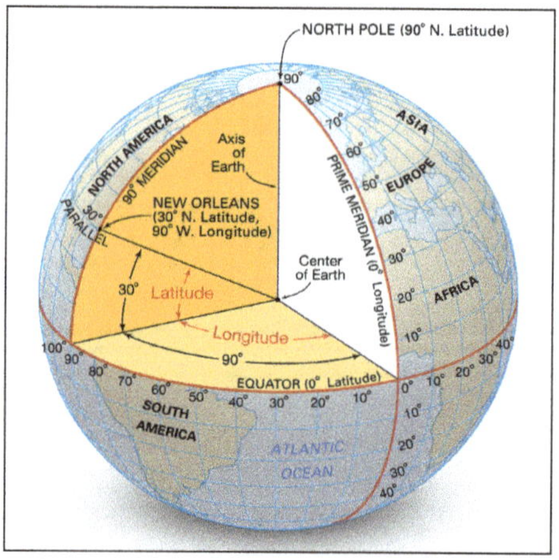

This cutaway drawing shows that the latitude and longitude of any place are based on the sizes of two angles that originate at the centre of the Earth.

Latitude is a measurement on a globe or map of location north or south of the Equator. Technically, there are different kinds of latitude—geocentric, astronomical, and geographic (or geodetic)—but there are only minor differences between them. In most common references, geocentric latitude is implied. Given in degrees, minutes, and seconds, geocentric latitude is the arc subtended by an angle at Earth's centre and measured in a north-south plane poleward from the Equator. Thus, a point at 30°15′20″ N subtends an angle of 30°15′20″at the centre of the globe; similarly, the arc between the

Equator and either geographic pole is 90° (one-fourth the circumference of the Earth, or $^{1}/_{4} \times 360°$), and thus the greatest possible latitudes are 90 °N and 90 °S. As aids to indicate different latitudinal positions on maps or globes, equidistant circles are plotted and drawn parallel to the Equator and each other; they are known as parallels, or parallels of latitude.

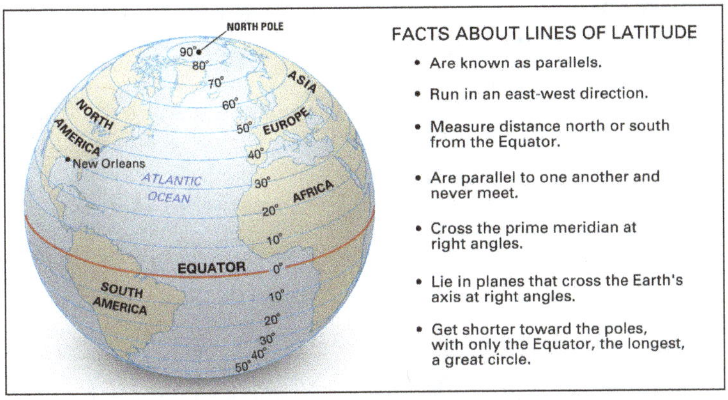

In contrast, geographic latitude, which is the kind used in mapping, is calculated using a slightly different process. Because Earth is not a perfect sphere—the planet's curvature is flatter at the poles—geographic latitude is the arc subtended by the equatorial plane and the normal line that can be drawn at a given point on Earth's surface. (The normal line is perpendicular to a tangent line touching Earth's curvature at that point on the surface.) Different methods are used to determine geographic latitude, as by taking angle-sights on certain polar stars or by measuring with a sextant the angle of the noon Sun above the horizon. The length of a degree of arc of latitude is approximately 111 km (69 miles), varying, because of the nonuniformity of Earth's curvature, from 110.567 km (68.706 miles) at the Equator to 111.699 km (69.41 miles) at the poles. Geographic latitude is also given in degrees, minutes, and seconds.

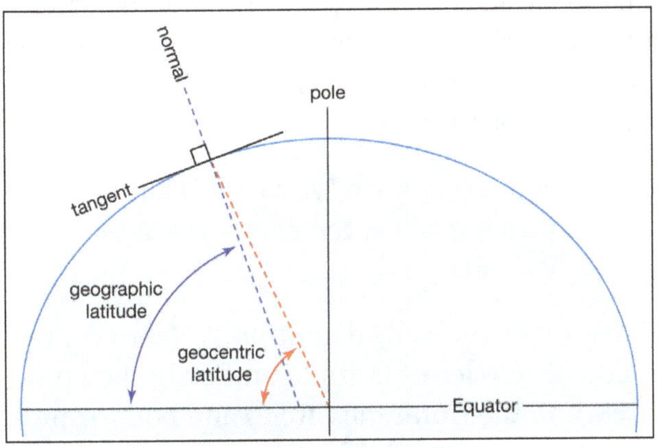

Latitude and longitude.

Longitude is a measurement of location east or west of the prime meridian at Greenwich, the specially designated imaginary north-south line that passes through both geographic poles and Greenwich, London. Measured also in degrees, minutes, and seconds, longitude is the amount of arc created by drawing first a line from the Earth's centre to the intersection of the Equator and the prime meridian and then another line from the Earth's centre to any point elsewhere on the Equator. Longitude is measured 180° both east and west of the prime meridian. As aids to locate longitudinal positions on a globe or map, meridians are plotted and drawn from pole to pole where they meet. The distance per degree of longitude at the Equator is about 111.32 km (69.18 miles) and at the poles, 0.

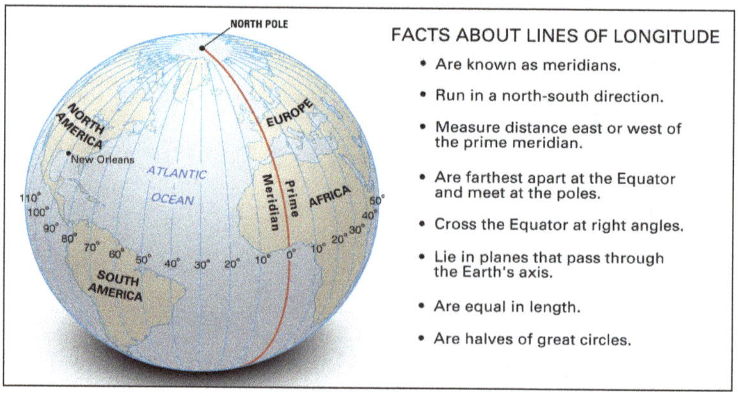

The combination of meridians of longitude and parallels of latitude establishes a framework or grid by means of which exact positions can be determined in reference to the prime meridian and the Equator: a point described as 40 °N, 30 °W, for example, is located 40° of arc north of the Equator and 30° of arc west of the Greenwich meridian.

EQUATOR

An equator of a rotating spheroid (such as a planet) is its zeroth circle of latitude (parallel). It is the imaginary line on the spheroid, equidistant from its poles, dividing it into northern and southern hemispheres. In other words, it is the intersection of the spheroid with the plane perpendicular to its axis of rotation and midway between its geographical poles.

On Earth, *the Equator* is about 40,075 km (24,901 mi) long, of which 78.8% lies across water and 21.3% over land. Indonesia is the country straddling the greatest length of the equatorial line across both land and sea.

The latitude of the Earth's equator is, by definition, 0° (zero degrees) of arc. The Equator is one of the five notable circles of latitude on Earth; the other four are both polar circles (the Arctic Circle and the Antarctic Circle) and both tropical circles (the Tropic of Cancer and the Tropic of Capricorn). The Equator is the only line of latitude which is also a great circle — that is, one whose plane passes through the center of the globe. The plane of Earth's equator, when projected outwards to the celestial sphere, defines the celestial equator.

Left: A monument marking the Equator near the city of Pontianak, Indonesia.
Right: Road sign marking the Equator near Nanyuki, Kenya.

In the cycle of Earth's seasons, the equatorial plane runs through the Sun twice per year: on the equinoxes in March and September. To a person on Earth, the Sun appears to travel above the Equator (or along the celestial equator) at these times. Light rays from the Sun's center are perpendicular to Earth's surface at the point of solar noon on the Equator.

The Equator marked as it crosses Ilhéu das Rolas, in São Tomé and Príncipe.

The Marco Zero monument marking the Equator in Macapá, Brazil.

Locations on the Equator experience the shortest sunrises and sunsets because the Sun's daily path is nearly perpendicular to the horizon for most of the year. The length of daylight (sunrise to sunset) is almost constant throughout the year; it is about 14 minutes longer than nighttime due to atmospheric refraction and the fact that sunrise begins (or sunset ends) as the upper limb, not the center, of the Sun's disk contacts the horizon.

Earth bulges slightly at the Equator; the "average" diameter of Earth is 12,750 km (7,920 mi), but the diameter at the Equator is about 43 km (27 mi) greater than at the poles.

Sites near the Equator, such as the Guiana Space Centre in Kourou, French Guiana, are good locations for spaceports as they have a fastest rotational speed of any latitude, 460 m/s. The added velocity reduces the fuel needed to launch spacecraft eastward (in the direction of Earth's rotation) to orbit, while simultaneously avoiding costly maneuvers to flatten inclination during missions such as the Apollo moon landings.

Geodesy

Precise Location

The precise location of the Equator is not truly fixed; the true equatorial plane is perpendicular to the Earth's spin axis, which drifts about 9 metres (30 ft) during a year. This effect must be accounted for in detailed geophysical measurements.

Exact Length

The International Association of Geodesy (IAG) and the International Astronomical Union (IAU) have chosen to use an equatorial radius of 6,378.1366 kilometres (3,963.1903 mi). This equatorial radius is also in the 2003 and 2010 IERS Conventions. It is also the equatorial radius used for the IERS 2003 ellipsoid. If it were really circular, the length of the Equator would then be exactly 2π times the radius, namely 40,075.0142 kilometres (24,901.4594 mi). The GRS 80 as approved and adopted by the IUGG at its Canberra, Australia meeting of 1979 has an equatorial radius of 6,378.137 kilometres (3,963.191 mi). The WGS 84 which is a standard for use in cartography, geodesy, and satellite navigation including GPS, also has an equatorial radius of 6,378.137 kilometres (3,963.191 mi). For both GRS 80 and WGS 84, this results in a length for the Equator of 40,075.0167 km (24,901.4609 mi).

The geographical mile is defined as one arc-minute of the Equator, so it has different values depending on which radius is assumed. For example, by WSG-84, the distance is 1,855.3248 metres (6,087.024 ft), while by IAU-2000, it is 1,855.3257 metres (6,087.027 ft). This is a difference of less than one millimetre (0.039 in) over the total distance (approximately 1.86 kilometres or 1.16 miles).

The earth is commonly modeled as a sphere flattened 0.336% along its axis. This makes the Equator 0.16% longer than a meridian (a great circle passing through the two poles). The IUGG standard meridian is, to the nearest millimetre, 40,007.862917 kilometres

(24,859.733480 mi), one arc-minute of which is 1,852.216 metres (6,076.82 ft), explaining the SI standardization of the nautical mile as 1,852 metres (6,076 ft), more than 3 metres (9.8 ft) less than the geographical mile.

The sea-level surface of the Earth (the geoid) is irregular, so the actual length of the Equator is not so easy to determine. *Aviation Week and Space Technology* on 9 October 1961 reported that measurements using the Transit IV-A satellite had shown the equatorial diameter from longitude 11° West to 169° East to be 1,000 feet (300 m) greater than its diameter ninety degrees away.

Equatorial Seasons and Climate

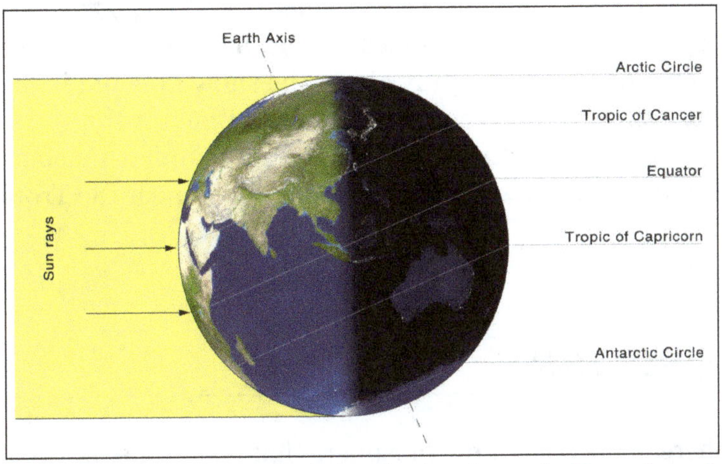

Above figure is of the seasons, depicting the situation at the December solstice.

Regardless of the time of day (i.e. the Earth's rotation on its axis), the North Pole will be dark, and the South Pole will be illuminated. In addition to the density of incident light, the dissipation of light in the atmosphere is greater when it falls at a shallow angle.

Seasons result from the tilt of the Earth's axis compared to the plane of its revolution around the Sun. Throughout the year the northern and southern hemispheres are alternately turned either toward or away from the sun depending on Earth's position in its orbit. The hemisphere turned toward the sun receives more sunlight and is in summer, while the other hemisphere receives less sun and is in winter.

At the equinoxes, the Earth's axis is perpendicular to the sun rather than tilted toward or away, meaning that day and night are both about 12 hours long across the whole of the Earth.

The Equator lies mostly on the three largest oceans: the Atlantic Ocean, the Indian Ocean, and the Pacific Ocean. Near the Equator there is little temperature change throughout the year, though there may be dramatic differences in rainfall and humidity. The terms summer, autumn, winter and spring do not generally apply. Lowlands around the Equator generally have a tropical rainforest climate, also known as an equatorial climate, though cold currents cause some regions to have tropical monsoon climates with a dry season in the middle of the year, and the Somali Current generated by the Asian monsoon due to continental heating via the high Tibetan Plateau causes Greater Somalia to have an arid climate despite its equatorial location.

Average annual temperatures in equatorial lowlands are around 31 °C (88 °F) during the afternoon and 23 °C (73 °F) around sunrise. Rainfall is very high away from cold current upwelling zones, from 2,500 to 3,500 mm (100 to 140 in) per year. There are about 200 rainy days per year and average annual sunshine hours are around 2,000. Despite high year-round sea level temperatures, some higher altitudes such as the Andes and Mount Kilimanjaro have glaciers. The highest point on the Equator is at the elevation of 4,690 metres (15,387 ft), at 0°0′0″N 77°59′31″W0.00000°N 77.99194°W, found on the southern slopes of Volcán Cayambe [summit 5,790 metres (18,996 ft)] in Ecuador. This is slightly above the snow line and is the only place on the Equator where snow lies on the ground. At the Equator the snow line is around 1,000 metres (3,300 ft) *lower* than on Mount Everest and as much as 2,000 metres (6,600 ft) lower than the highest snow line in the world, near the Tropic of Capricorn on Llullaillaco.

DEEP BIOSPHERE

The deep biosphere is the part of the biosphere that resides below the first few meters of the surface. It extends down at least 5 kilometers below the continental surface and 10.5 kilometers below the sea surface. It includes all three domains of life and the genetic diversity rivals that on the surface.

The deep biosphere is life found in the deep subsurface. For the seafloor, an operational definition of deep subsurface is the region that is not penetrated by seawater or bioturbated by animals; this is generally about a meter or more below the surface. On continents, it is below a few meters, not including soils. The organisms in this zone are sometimes referred to as intraterrestrials.

Scientific Methods

The present understanding of subsurface biology was made possible by numerous advances in technology for sample collection, field analysis, molecular science, cultivation, imaging and analysis.

Sample Collection

The ocean floor is sampled by drilling boreholes and collecting cores. The methods must be adapted to different types of rock, and the cost of drilling limits the number of holes that can be drilled. Microbiologists have made use of scientific drilling programs: the Ocean Drilling Program (ODP), which used the JOIDES Resolution drilling platform, and the Integrated Ocean Drilling Program (IODP), which used the Japanese ship Chikyū.

Chikyu 1 drilling ship.

Other approaches have been used, particularly the use of deep mines. These mines have been excavated to extract resources, for example the South African gold mines, or chosen as potential nuclear waste repositories (e.g., Yucca Mountain and the Waste Isolation Pilot Plant in the United States, Äspö in Sweden), so they are likely to present a biased view of the underground life.

To allow continuous underground sampling, various kinds of observatories have been developed. On the ocean floor, the Circulation Obviation Retrofit Kit (CORK) seals a borehole to cut off the influx of seawater, while the Simple Cabled Instrument for Measuring Parameters In-Situ (SCIMPI) functions in sediments after a borehole has collapsed. Similarly, devices such as the flow-through *in situ* reactor (FTISR) are used to sample the continental subsurface. Various methods are used to extract fluids from these sites, including passive gas samplers, U-tube systems and osmotic gas samplers.

Field Analysis and Manipulation

Some methods analyze microbes *in situ* rather than extract them. In biogeophysics, the effects of microbes on properties of geological materials are remotely probed using electrical signals. Microbes can be tagged using a stable isotope such as carbon-13 and then re-injected in the ground to see where they go. A "push-pull" method involves injection of a fluid into an aquifer and extraction of a mixture of injected fluid with the ground water; the latter can then be analyzed to determine what chemical reactions occurred.

Molecular Methods and Cultivation

Methods from modern medicine allow the extraction of nucleic acids, lipids and proteins from cells, DNA sequencing, and the physical and chemical analysis of molecules using mass spectrometry and flow cytometry. A lot can be learned about the microbial communities using these methods even when the individuals cannot be cultivated. For example, at the Richmond Mine in California, scientists used shotgun sequencing to identify four new species of bacteria, three new species of archaea (known as the Archaeal Richmond Mine acidophilic nanoorganisms), and 572 proteins unique to the bacteria.

Extent

Life has been found at depths of 5 km in continents and 10.5 km below the ocean surface. The estimated volume of the deep biosphere is 2–2.3 billion cubic kilometers, about twice the volume of the oceans.

Diversity

The biomass of the subsurface life is about 15% of the total for the biosphere. Life from all three domains of life (Archaea, Bacteria, and Eukarya) have been found in the deep subsurface. The genetic diversity is at least as great as that on the surface. Bacteria are mainly Proteobacteria while the Archaea are mainly Methanomicrobia and Thaumarchaeota. The subsurface accounts for about 90% of all the biomass in Archaea and Bacteria. One species of bacterium, "*Candidatus* Desulforudis audaxviator", is the first known to comprise a complete ecosystem by itself.

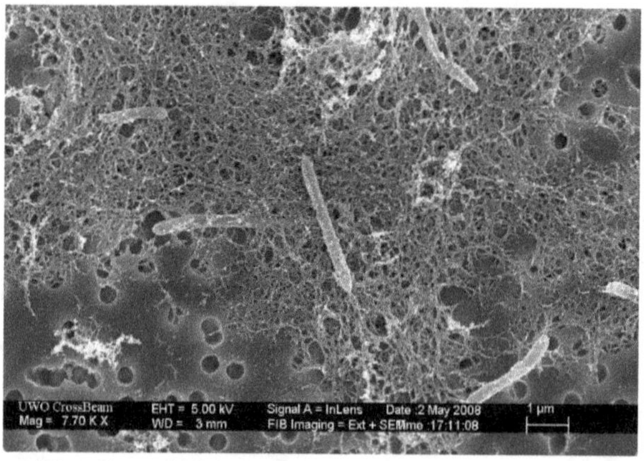

Ca. Desulforudis audaxviator from a gold mine near Johannesburg.

The eukaryotes include some multicellular life. In 2009 a species of nematode, *Halicephalobus mephisto*, was discovered in rock fissures more than a kilometer down a South African gold mine. Nicknamed the "devil worm", it may have been forced down along with pore water by earthquakes. Other multicellular organisms have since been found, including fungi, Platyhelminthes (flatworms), Rotifera, Annelida (ringed worms) and Arthropoda.

ARCTIC CIRCLE

The Arctic Circle is one of the two polar circles and the most northerly of the five major circles of latitude as shown on maps of Earth. It marks the northernmost point at which the centre of the noon sun is just visible on the December solstice and the southernmost point at which the centre of the midnight sun is just visible on the June solstice. The region north of this circle is known as the Arctic, and the zone just to the south is called the Northern Temperate Zone.

As seen from the Arctic, the Sun is above the horizon for 24 continuous hours at least once per year (and therefore visible at midnight) and below the horizon for 24 continuous hours at least once per year (and therefore not visible at noon). This is also true in the Antarctic region, south of the equivalent Antarctic Circle.

The position of the Arctic Circle is not fixed; as of 20 August 2019, it runs 66° 33′47.7″ north of the Equator. Its latitude depends on the Earth's axial tilt, which fluctuates within a margin of more than 2° over a 41,000-year period, due to tidal forces resulting from the orbit of the Moon. Consequently, the Arctic Circle is currently drifting northwards at a speed of about 15 m (49 ft) per year.

Midnight Sun and Polar Night

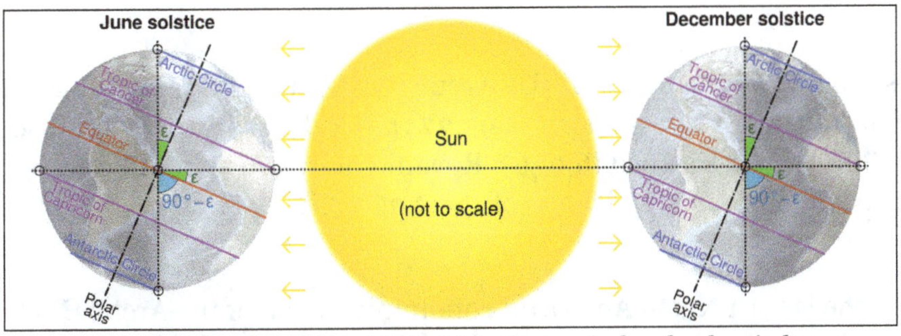

Relationship of Earth's axial tilt (ε) to the tropical and polar circles.

The Arctic Circle is the southernmost latitude in the Northern Hemisphere at which the centre of the sun can remain continuously above or below the horizon for twenty-four hours; as a result, at least once each year at any location within the Arctic Circle the sun is visible at local midnight, and at least once the centre is not visible at local noon.

Directly on the Arctic Circle these events occur, in principle, exactly once per year: at the June and December solstices, respectively. However, because of atmospheric refraction and mirages, and also because the sun appears as a disk and not a point, part of the midnight sun may be seen on the night of the northern summer solstice up to about 50 minutes (′) (90 km (56 mi)) south of the Arctic Circle; similarly, on the day of the northern winter solstice, part of the sun may be seen up to about 50′ north of the Arctic Circle. That is true at sea level; those limits increase with elevation above sea level, although in mountainous regions there is often no direct view of the true horizon.

Human Habitation

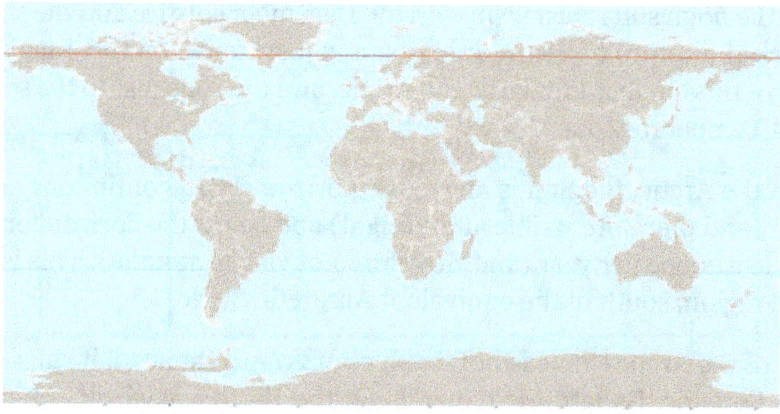

Cylindrical projection showing the Arctic Circle in red.

Only four million people live north of the Arctic Circle due to the climate; nonetheless, some areas have been settled for thousands of years by indigenous peoples, who today make up 10% of the region's population. Tens of thousands of years ago, waves of people migrated from eastern Siberia across the Bering Strait into North America to settle.

The largest communities north of the Arctic Circle are situated in Russia, Norway, and Sweden: Murmansk (population 295,374), Norilsk (178,018), Tromsø (75,638), Vorkuta (58,133), and Kiruna (16,936). Rovaniemi (62,667) in Finland is the largest settlement in the immediate vicinity of the Arctic Circle, lying 6 km (4 mi) south of the line.

In contrast, the largest North American community north of the Arctic Circle, Sisimiut (Greenland), has approximately 5,000 inhabitants. Of the Arctic communities in Canada and the United States, Utqiagvik, Alaska is the largest settlement with about 4,000 inhabitants.

Geography

The Arctic Circle is roughly 16,000 km (9,900 mi) long. The area north of the Circle is about 20,000,000 km^2 (7,700,000 sq mi) and covers roughly 4% of Earth's surface.

The Arctic Circle passes through the Arctic Ocean, the Scandinavian Peninsula, North Asia, Northern America, and Greenland. The land within the Arctic Circle is divided among eight countries: Norway, Sweden, Finland, Russia, the United States (Alaska), Canada (Yukon, Northwest Territories, and Nunavut), Denmark (Greenland), and Iceland (where it passes through the small offshore island of Grímsey).

Climate

The climate inside the Arctic Circle is generally cold, but the coastal areas of Norway have a generally mild climate as a result of the Gulf Stream, which makes the ports of northern Norway and northwest Russia ice-free all year long. In the interior, summers can be quite warm, while winters are extremely cold. For example, summer temperatures in Norilsk, Russia will sometimes reach as high as 30 °C (86 °F), while the winter temperatures frequently fall below −50 °C (−58 °F).

ANTARCTIC CIRCLE

The Antarctic Circle is one of the five major circles or parallels of latitude that mark maps of the Earth.

It is at latitude 66 degrees 33′ 39″ south of the equator (in 2000; like its northern counterpart, the Arctic Circle, the value is currently slowly decreasing over time, pushing the Antarctic Circle southwards with about 15 m per year).

For everywhere within the Antarctic Circle, there is at least twenty-four hours of continuous daylight on the Summer Solstice in December, and at least twenty-four hours of continuous nighttime on the Winter Solstice in June.

That is to say, one whole day during which the sun does not set, and one whole day during which the sun does not rise.

This is because the earth is tilted at a 23.5 degree angle, and during the winter solstice, the southern hemisphere is tilted away from the Sun, meaning that the antarctic circle is completely tilted away from the Sun, hence it experiences 24 hour nighttime, and vice versa.

References

- Latitude, science: britannica.com, Retrieved 2 June, 2019
- Hoehler, tori m.; jorgensen, bo barker (16 january 2013). "microbial life under extreme energy limitation". Nature reviews microbiology. 11: 83–94. Doi:10.1038/nrmicro2939
- Schippers, a. (2015). "deep biosphere". In harff, j.; meschede, m.; petersen, s.; thiede, j. (eds.). Encyclopedia of marine geosciences. Dordrecht: springer. Isbn 978-94-007-6644-0
- Edwards, katrina (2 september 2011). "north pond: searching for intraterrestrial life". Scientific american blog network. Retrieved 18 january2019
- Marsh, william m.; kaufman, martin m. (2012). Physical geography: great systems and global environments. Cambridge university press. P. 24. Isbn 978-0-521-76428-5
- Antarctic-circle, terms: sciencedaily.com, Retrieved 12 June, 2019

5

Flora and Fauna in the Biosphere

Flora refers to the entirety of plants found in a particular region. Fauna is a term used to represent animal life which is indigenous to a specific region. This chapter has been carefully written to provide an extensive understanding of the importance of flora and fauna in the biosphere.

FLORA

Flora is the name given to the collective plant life that grows or once grew in a certain area or during a given time period. It usually refers to the native plant life present but does include new species that have been introduced as well.

The study of the plant life around the world is very interesting since is makes different classifications of the flora.

Plant life on the earth can be distinguished from each other in a number of ways. The simplest is to divide on the basis of region. Plants that grow specifically in the mountains will be very different from those that grow in the desert. Similarly, the plants that have adapted to living underwater are treated as a unique form of flora. Scientists can also study 'Fossil Flora', which comprises of plant life that was found in pre-historic times. The current flora and fauna of the earth is also divided on the basis of the environment in which it is grown or seen naturally.

When we talk about 'Native Flora', we are referring to the plant life that is indigenous to a particular region. Cacti are the native flora of deserts all over the world. They can grow in most weather conditions but are native to the sandy dunes of deserts. When we refer to 'Agricultural Flora', we are talking about plant life that has been grown by humans for a certain purpose.

They may or may not be native, but are used by humans for their own needs. Similarly there is 'Garden Flora' or 'Horticultural Flora', which are plants grown for decorative purposes. And then we have the 'Weed Flora', which are plants that are either undesirable in certain areas or invasive within the native plant life.

Manchineel Tree

It's a flora that falls under the species of flowering plants in the spurge family. It originates in northern South America and southern North America. It bears grapes that can result in rashes if you're wet. It's not advisable to shelter under it when it rains since it produces milk that causes burns and blisters. Also, desist from eating its fruits as they cause the same effects to your mouth and throat.

Giant Redwood Trees

We are used to the typical trees growing in our city parks or neighborhoods. Some species of fauna exist that grow to astronomical heights. The Giant redwood tree is a classic example. This kind of tree is considered the largest and tallest on earth. In fact, one specimen registered well above 350 feet tall.

Giant redwood trees are native to the north coast of California. The conditions here that include humid and rainy weather encourage such astronomical growths. The nature of these tree species growth allows roads near the Sequoia National Park to snake through them. The canopy of these trees provides home for a wide range of bird species. The forest below harbors most animal species.

Nepenthes

Giant redwood trees and Machineel tree are photosynthetic; which means they manufacture their own food from the sun rays. However, other kinds of flora take a lot more proactive approach to developing their own food. Nepenthe is a typical example. It's classified under the family of pitcher plants and originates from the jungles of Borneo. These floras have adapted leaves, which form a huge pitcher structure. The pitcher contains sweet smelling liquid meant to attract insects and even some minute frog species. The insects and frogs drop inside the pitcher, but coming out becomes impossible due to the smooth, slippery sides of the pitcher. The pitcher also contains digestive juices that immediately begin to act and digest the insects and frogs, enabling the plant to benefit from nutrients not present in the soil.

Welwitschia Mirabilis

Welwitschia Mirabilis is one plant species that doesn't like rain. It's native to the Namib Desert and thrives in the driest conditions on the planet. Oddly enough, it only features 2 large leaves that wrap around it depicting a wilted or dry plant. Despite its less attractive nature, this plant species is capable of living for more than 1500 years. This is possible due to its characteristic deep root system that can tap underground water even in the driest conditions.

Horticultural Flora

A horticultural flora, also known as a garden flora, is a plant identification aid structured in the same way as a native plants flora. It serves the same purpose: to facilitate plant identification; however, it only includes plants that are under cultivation as ornamental plants growing within the prescribed climate zone or region.

Scope

Horticultural floras include both cultigens (plants deliberately altered in some way by human activity) and those wild plants brought directly into cultivation that do not have cultigen names. They might also include colour images and useful information specific to the zone or region including:

- Historical details about outstanding public and private cultivated plant collections.
- Exceptional trees (age, history, rarity, size etc.).
- Prominent nurserymen and plant breeders.
- References to the taxonomic and other literature on the plant groups.
- Easy "spotting" or "field" characters useful for quick identification.
- Notes on ecology (especially the potential of plants to naturalise and become invasive).
- Horticultural history of introduction.
- Conservation.

Uses

Written by a professional plant taxonomist or plantsperson, a horticultural flora assists clarification of scientific and common names, the identification of plant characteristics that occur in cultivated plants that are additional to those in wild counterparts, and descriptions of cultivars.

Although horticultural floras may include a range of food plants, their emphasis is generally on ornamental plants and so these floras are sometimes referred to as "garden floras". Increasingly they provide data for sustainable landscaping, such as:

- Drought tolerance and irrigation needs.
- Food sources of native and migratory bird and butterfly species.
- Companion planting.

- Invasive species notation.
- Local habitat restoration aspects.

FAUNA

Fauna is the name given to collective animal life that lives or was once found in a certain area or time period.

Both the flora and fauna of the earth have been given sub-divisions. Fauna is also distinguished in many different ways. However, these ways are much more complex than floral divisions because animal life has evolved into many different forms.

To begin with, the animal kingdom by itself is Fauna. However, within it we have the 'Avifauna' that refers to the birds and 'Piscifauna' referring to the fish. These are simpler classifications, since they do not cover small single celled organisms such as bacteria and virus. They also do not account for the microscopic organisms that abound in nature but cannot be seen by the human eye. Such animal life is known as 'Microfauna'. Much of the flora and fauna of the earth can be seen, but there is a very large percentage that has yet to be named and even discovered.

This is why classifications such as 'Cryptofauna' arise, which refers to animals that are extremely rare and may even be mythological. There are also 'Microfauna', which refers to the extremely small animals and 'Megafauna', that talks about the large animals we usually see.

Planet earth has been blessed in more ways than one. We have complex ecological systems that not only provide us with resources, but make the world a more beautiful place to live in.

Coral

Apart from typical furry mammals such as bears and wolves, aquatic animals like corals, most insects on land and jellyfish make up the fauna. Coral reefs are the eye-catching colored landscapes beneath the ocean surface. On sight, you might think they are plants, but they are animal species constituting the fauna of the ocean. Corals are immobile animals, and so they feed by picking out prey that swims by, for example, the zooplankton. An example of a coral is the Great Barrier Reef in Australia, which is the largest natural coral on the globe. It harbors thousand of flora and fauna in warm shallow water in proximity to the equator.

Southern Toadlet

This animal species features an olive to dark brown body and wrapped with many small warts over its body. It also has a black and white marbled look around its belly. The Southern Toadlet thrives in damp habitats such as drainage lines, woodlands, scrubland, dry forests, and grassland. This animal species is native to the south-east of South Australia, south Victoria, and eastern Tasmania.

Eltham Copper Butterfly

It is an endangered species of butterfly. Eltham area in Victoria supports a greater population of this kind of butterfly. This butterfly species was initially discovered near Eltham in 1938. It was later thought to be extinct in the 1950s. The Eltham Copper butterfly is small yet glittering species of butterfly coupled with bright copper coloring above its wings. The copper coloring makes it clearly visible in the course of summer flight season.

There are hundreds of thousands of flora and fauna species across the globe. Flora and fauna bring out the true beauty of planet earth and enables humans to experience two separate worlds in one.

Megafauna

In terrestrial zoology, megafauna are large or giant animals. The most common thresholds used are weight over 40 kilograms (90 lb) or 44 kilograms (100 lb) (i.e., comparable or larger in mass than a human) or over a tonne, 1,000 kilograms (2,205 lb) (i.e., having a mass comparable to or larger than an ox). The first of these include many species not popularly thought of as overly large, such as white-tailed deer and red kangaroo.

In practice, the most common usage encountered in academic and popular writing describes land mammals roughly larger than a human that are not (solely) domesticated. The term is especially associated with the Pleistocene megafauna – the land animals often larger than modern counterparts considered archetypical of the last ice age, such as mammoths, the majority of which in northern Eurasia, the Americas and Australia became extinct within the last forty thousand years. It is also commonly used for the largest extant wild land animals, especially elephants, giraffes, hippopotami, rhinoceroses, and large bovines. (Of these five categories of large herbivorous mammals, only bovines are presently found outside of Africa and southern Asia, but all the others were formerly more wide-ranging.) Megafaunal species may be subcategorized by their trophic position into megaherbivores (e.g., elephants), megacarnivores (e.g., lions), and, more rarely, megaomnivores (e.g., bears).

Other common uses are for giant aquatic species, especially whales, any larger wild or domesticated land animals such as larger antelope and cattle, as well as numerous dinosaurs and other extinct giant reptilians.

The term is also sometimes applied to animals (usually extinct) of great size *relative to* a more common or surviving type of the animal, for example the 1 m (3 ft) dragonflies of the Carboniferous period.

Ecological Strategy

Megafauna – in the sense of the largest mammals and birds – are generally K-strategists, with high longevity, slow population growth rates, low mortality rates, and (at least for the largest) few or no natural predators capable of killing adults. These characteristics, although not exclusive to such megafauna, make them vulnerable to human overexploitation, in part because of their slow population recovery rates.

Evolution of Large Body Size

One observation that has been made about the evolution of larger body size is that rapid rates of increase that are often seen over relatively short time intervals are not sustainable over much longer time periods. In an examination of mammal body mass changes over time, the maximum increase possible in a given time interval was found to scale with the interval length raised to the 0.25 power. This is thought to reflect the emergence, during a trend of increasing maximum body size, of a series of anatomical, physiological, environmental, genetic and other constraints that must be overcome by evolutionary innovations before further size increases are possible. A strikingly faster rate of change was found for large decreases in body mass, such as may be associated with the phenomenon of insular dwarfism. When normalized to generation length, the maximum rate of body mass decrease was found to be over 30 times greater than the maximum rate of body mass increase for a ten-fold change.

In Terrestrial Mammals

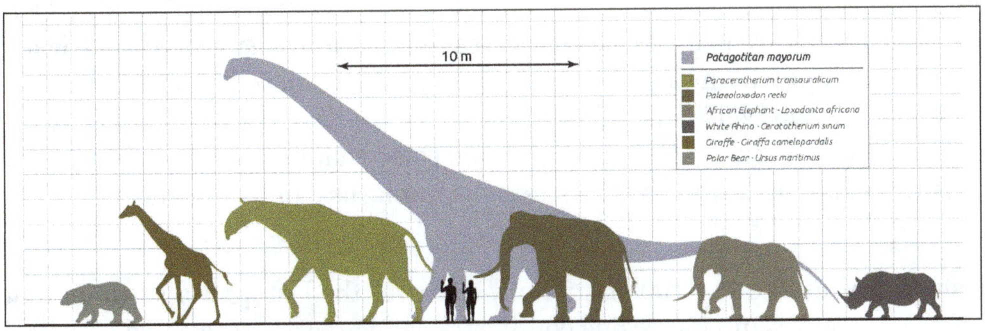

Large terrestrial mammals compared in size to one of the largest sauropod dinosaurs, Patagotitan.

Subsequent to the Cretaceous–Paleogene extinction event that eliminated the non-avian dinosaurs about 66 Ma (million years) ago, terrestrial mammals underwent a nearly exponential increase in body size as they diversified to occupy the ecological niches left vacant. Starting from just a few kg before the event, maximum size had reached ~50 kg a few million years later, and ~750 kg by the end of the

Paleocene. This trend of increasing body mass appears to level off about 40 Ma ago (in the late Eocene), suggesting that physiological or ecological constraints had been reached, after an increase in body mass of over three orders of magnitude. However, when considered from the standpoint of rate of size increase per generation, the exponential increase is found to have continued until the appearance of *Indricotherium* 30 Ma ago. (Since generation time scales with *body mass*, increasing generation times with increasing size cause the log mass vs. time plot to curve downward from a linear fit.)

Megaherbivores eventually attained a body mass of over 10,000 kg. The largest of these, indricotheres and proboscids, have been hindgut fermenters, which are believed to have an advantage over foregut fermenters in terms of being able to accelerate gastrointestinal transit in order to accommodate very large food intakes. A similar trend emerges when rates of increase of maximum body mass per generation for different mammalian clades are compared (using rates averaged over macroevolutionary time scales). Among terrestrial mammals, the fastest rates of increase of *body mass* vs. time (in Ma) occurred in perissodactyls (a slope of 2.1), followed by rodents (1.2) and proboscids (1.1), all of which are hindgut fermenters. The rate of increase for artiodactyls (0.74) was about a third that of perissodactyls. The rate for carnivorans (0.65) was slightly lower yet, while primates, perhaps constrained by their arboreal habits, had the lowest rate (0.39) among the mammalian groups studied.

Terrestrial mammalian carnivores from several eutherian groups (the artiodactyl *Andrewsarchus* - formerly considered a mesonychid, the oxyaenid *Sarkastodon*, and the carnivorans *Amphicyon* and *Arctodus*) all reached a maximum size of about 1000 kg (the carnivoran *Arctotherium* and the hyaenodontid *Simbakubwa* may have been somewhat larger). The largest known metatherian carnivore, *Proborhyaena gigantea*, apparently reached 600 kg, also close to this limit. A similar theoretical maximum size for mammalian carnivores has been predicted based on the metabolic rate of mammals, the energetic cost of obtaining prey, and the maximum estimated rate coefficient of prey intake. It has also been suggested that maximum size for mammalian carnivores is constrained by the stress the humerus can withstand at top running speed.

Analysis of the variation of maximum body size over the last 40 Ma suggests that decreasing temperature and increasing continental land area are associated with increasing maximum body size. The former correlation would be consistent with Bergmann's rule, and might be related to the thermoregulatory advantage of large body mass in cool climates, better ability of larger organisms to cope with seasonality in food supply, or other factors; the latter correlation could be explained in terms of range and resource limitations. However, the two parameters are interrelated (due to sea level drops accompanying increased glaciation), making the driver of the trends in maximum size more difficult to identify.

In Marine Mammals

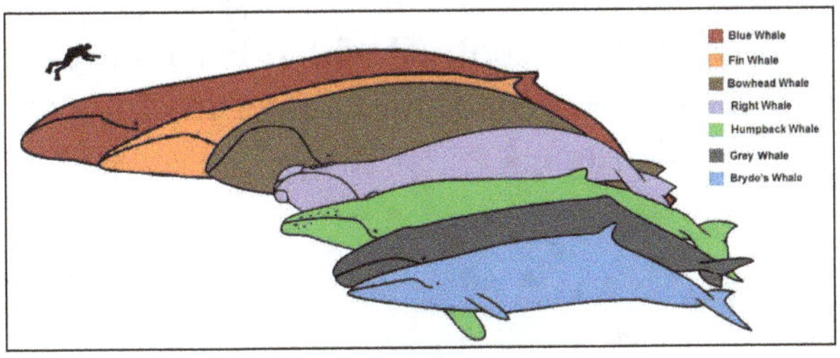

Baleen whale comparative sizes.

Since tetrapods (first reptiles, later mammals) returned to the sea in the Late Permian, they have dominated the top end of the marine body size range, due to the more efficient intake of oxygen possible using lungs. The ancestors of cetaceans are believed to have been the semiaquatic pakicetids, no larger than wolves, of about 53 million years (Mya) ago. By 40 Ma ago, cetaceans had attained a length of 20 m or more in *Basilosaurus*, an elongated, serpentine whale that differed from modern whales in many respects and was not ancestral to them. Following this, the evolution of large body size in cetaceans appears to have come to a temporary halt, and then to have backtracked, although the available fossil records are limited. However, in the period from 31 Ma ago (in the Oligocene) to the present, cetaceans underwent a significantly more rapid sustained increase in body mass (a rate of increase in *body mass* of a factor of 3.2 per million years) than achieved by any group of terrestrial mammals. This trend led to the largest animal of all time, the modern blue whale. Several reasons for the more rapid evolution of large body size in cetaceans are possible. Fewer biomechanical constraints on increases in body size may be associated with suspension in water as opposed to standing against the force of gravity, and with swimming movements as opposed to terrestrial locomotion. Also, the greater heat capacity and thermal conductivity of water compared to air may increase the thermoregulatory advantage of large body size in marine endotherms, although diminishing returns apply.

Cetaceans are not the only marine mammals to reach unprecedented size in the modern era. The largest carnivoran of all time is the mostly aquatic modern southern elephant seal.

In Flightless Birds

Because of the small initial size of all mammals following the extinction of the non-avian dinosaurs, nonmammalian vertebrates had a roughly ten-million-year-long window of opportunity (during the Paleocene) for evolution of gigantism without much competition. During this interval, apex predator niches were often occupied by reptiles, such as terrestrial crocodilians (e.g. *Pristichampsus*), large snakes (e.g. *Titanoboa*) or varanid lizards, or by flightless birds (e.g. *Paleopsilopterus* in South America). This is also the period when megafaunal flightless herbivorous gastornithid birds evolved in the Northern Hemisphere, while flightless paleognaths evolved to large size on Gondwanan land masses and Europe. Gastornithids and at least one lineage of flightless paleognath birds originated in Europe, both lineages dominating niches for large herbivores while mammals remained below 45 kg (in contrast with other landmasses like North America and Asia, which saw the earlier evolution of larger mammals) and were the largest European tetrapods in the Paleocene.

A size comparison between a human and 4 moa species: 1. Dinornis novaezealandiae 2. Emeus crassus 3. Anomalopteryx didiformis 4. Dinornis robustus.

Flightless paleognaths, termed ratites, have traditionally been viewed as representing a lineage separate from that of their small flighted relatives, the Neotropic tinamous. However, recent genetic studies have found that tinamous nest well within the ratite tree, and are the sister group of the extinct moa of New Zealand. Similarly, the small kiwi of New Zealand have been found to be the sister group of the extinct elephant birds of Madagascar. These findings indicate that flightlessness and gigantism arose independently multiple times among ratites via parallel evolution.

Predatory megafaunal flightless birds were often able to compete with mammals in the early Cenozoic. Later in the Cenozoic, however, they were displaced by advanced carnivorans and died out. In North America, the bathornithids *Paracrax* and *Bathornis* were apex predators but became extinct by the Early Miocene. In South America, the related phorusrhacids shared the dominant predatory niches with metatherian sparassodonts during most of the Cenozoic but declined and ultimately went extinct after eutherian predators arrived from North America (as part of the Great American Interchange) during the Pliocene. In contrast, large herbivorous flightless ratites have survived to the present.

However, none of the largest predatory (*Brontornis*), possibly omnivorous (*Dromornis*) or herbivorous (*Aepyornis*) flightless birds of the Cenozoic ever grew to masses much above 500 kg, and thus never attained the size of the largest mammalian carnivores, let alone that of the largest mammalian herbivores. It has been suggested that the increasing thickness of avian eggshells in proportion to egg mass with increasing egg size places an upper limit on the size of birds. The largest species of *Dromornis*, *D. stirtoni*, may have gone extinct after it attained the maximum avian body mass and was then outcompeted by marsupial diprotodonts that evolved to sizes several times larger.

In Giant Turtles

Giant tortoises were very important components of late Cenozoic faunas, occurring in virtually every continent until the arrival of homininans. The largest known terrestrial tortoise was *Megalochelys atlas*, an animal that probably weighed about 1,000 kg.

Some earlier aquatic Testudines, e.g. the marine *Archelon* of the Cretaceous and freshwater *Stupendemys* of the Miocene, were considerably larger, weighing more than 2,000 kg.

Megafaunal Mass Extinctions

Timing and Possible Causes

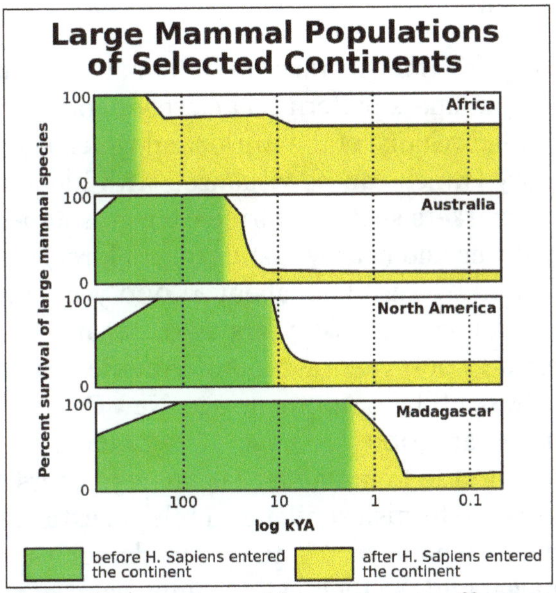

Correlations between times of first appearance of humans and unique megafaunal extinction pulses on different land masses.

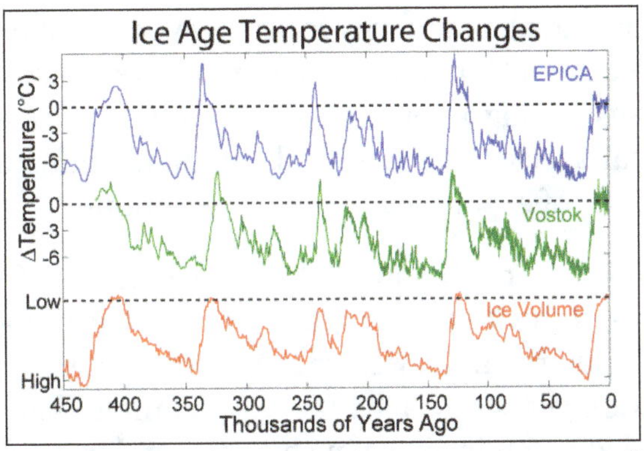

Cyclical pattern of global climate change over the last 450,000 years (based on Antarctic temperatures and global ice volume), showing that there were no unique climatic events that would account for any of the megafaunal extinction pulses.

The Holocene extinction, occurred at the end of the last ice age glacial period (a.k.a. the Würm glaciation) when many giant ice age mammals, such as woolly mammoths, went extinct in the Americas and northern Eurasia. An analysis of the extinction event in North America found it to be unique among Cenozoic extinction pulses in its selectivity for large animals. Various theories have attributed the wave of extinctions to human hunting, climate change, disease, a putative extraterrestrial impact, or other causes. However, this extinction near the end of the Pleistocene was just one of a series of megafaunal extinction pulses that have occurred during the last 50,000 years over much of the Earth's surface, with Africa and southern Asia (where the local megafauna had a chance to evolve alongside modern humans) being comparatively less affected. The latter areas did suffer a gradual attrition of megafauna, particularly of the slower-moving species (a class of vulnerable megafauna epitomized by giant tortoises), over the last several million years.

Outside the mainland of Afro-Eurasia, these megafaunal extinctions followed a highly distinctive landmass-by-landmass pattern that closely parallels the spread of humans into previously uninhabited regions of the world, and which shows no overall correlation with climatic history (which can be visualized with plots over recent geological time periods of climate markers such as marine oxygen isotopes or atmospheric carbon dioxide levels). Australia and nearby islands (e.g., Flores) were struck first around 46,000 years ago, followed by Tasmania about 41,000 years ago (after formation of a land bridge to Australia about 43,000 years ago), Japan apparently about 30,000 years ago, North America 13,000 years ago, South America about 500 years later, Cyprus 10,000 years ago, the Antilles 6,000 years ago, New Caledonia and nearby islands 3,000 years ago, Madagascar 2,000 years ago, New Zealand 700 years ago, the Mascarenes 400 years ago, and the Commander Islands 250 years ago. Nearly all of the world's isolated islands could furnish similar examples of extinctions occurring shortly after the arrival of humans, though most of these islands, such as the Hawaiian Islands, never had terrestrial megafauna, so their extinct fauna were smaller.

An analysis of the timing of Holarctic megafaunal extinctions and extirpations over the last 56,000 years has revealed a tendency for such events to cluster within interstadials, periods of abrupt warming, but only when humans were also present. Humans may have impeded processes of migration and recolonization that would otherwise have allowed the megafaunal species to adapt to the climate shift. In at least some areas, interstadials were periods of expanding human populations.

An analysis of *Sporormiella* fungal spores (which derive mainly from the dung of megaherbivores) in swamp sediment cores spanning the last 130,000 years from Lynch's Crater in Queensland, Australia, showed that the megafauna of that region virtually disappeared about 41,000 years ago, at a time when climate changes were minimal; the change was accompanied by an increase in charcoal, and was followed by a transition from rainforest to fire-tolerant sclerophyll vegetation. The high-resolution chronology of the changes supports the hypothesis that human hunting alone eliminated the megafauna, and that the subsequent change in flora was most likely a consequence of the elimination of browsers and an increase in fire. The increase in fire lagged the disappearance of megafauna by about a century, and most likely resulted from accumulation of fuel once browsing stopped. Over the next several centuries grass increased; sclerophyll vegetation increased with a lag of another century, and a sclerophyll forest developed after about another thousand years. During two periods of climate change about 120,000 and 75,000 years ago, sclerophyll vegetation had also increased at the site in response to a shift to cooler, drier conditions; neither of these episodes had a significant impact on megafaunal abundance. Similar conclusions regarding the culpability of human hunters in the disappearance of Pleistocene megafauna were derived from high-resolution chronologies obtained via an analysis of a large collection of eggshell fragments of the flightless Australian bird *Genyornis newtoni*, from analysis of *Sporormiella* fungal spores from a lake in eastern North America and from study of deposits of Shasta ground sloth dung left in over half a dozen caves in the American southwest.

Continuing human hunting and environmental disturbance has led to additional megafaunal extinctions in the recent past, and has created a serious danger of further extinctions in the near future. Direct killing by humans, primarily for meat, is the most significant factor in contemporary megafaunal decline.

A number of other mass extinctions occurred earlier in Earth's geologic history, in which some or all of the megafauna of the time also died out. Famously, in the Cretaceous–Paleogene extinction event the non-avian dinosaurs and most other giant reptilians were eliminated. However, the earlier mass extinctions were more global and not so selective for megafauna; i.e., many species of other types, including plants, marine invertebrates and plankton, went extinct as well. Thus, the earlier events must have been caused by more generalized types of disturbances to the biosphere.

Consequences of Depletion of Megafauna

Effect on Nutrient Transport

Megafauna play a significant role in the lateral transport of mineral nutrients in an ecosystem, tending to translocate them from areas of high to those of lower abundance. They do so by their movement between the time they consume the nutrient and the time they release it through elimination (or, to a much lesser extent, through decomposition after death). In South America's Amazon Basin, it is estimated that such lateral diffusion was reduced over 98% following the megafaunal extinctions that occurred roughly 12,500 years ago. Given that phosphorus availability is thought to limit productivity in much of the region, the decrease in its transport from the western part of the basin and from floodplains (both of which derive their supply from the uplift of the Andes) to other areas is thought to have significantly impacted the region's ecology, and the effects may not yet have reached their limits.

Effect on Methane Emissions

Large populations of megaherbivores have the potential to contribute greatly to the atmospheric concentration of methane, which is an important greenhouse gas. Modern ruminant herbivores produce methane as a byproduct of foregut fermentation in digestion, and release it through belching or flatulence. Today, around 20% of annual methane emissions come from livestock methane release. In the Mesozoic, it has been estimated that sauropods could have emitted 520 million tons of methane to the atmosphere annually, contributing to the warmer climate of the time (up to 10 °C warmer than at present). This large emission follows from the enormous estimated biomass of sauropods, and because methane production of individual herbivores is believed to be almost proportional to their mass.

Recent studies have indicated that the extinction of megafaunal herbivores may have caused a reduction in atmospheric methane. This hypothesis is relatively new. One study examined the methane emissions from the bison that occupied the Great Plains of North America before contact with European settlers. The study estimated that the removal of the bison caused a decrease of as much as 2.2 million tons per year. Another study examined the change in the methane concentration in the atmosphere at the end of the Pleistocene epoch after the extinction of megafauna in the Americas. After early humans migrated to the Americas about 13,000 BP, their hunting and other associated ecological impacts led to the extinction of many megafaunal species there. Calculations suggest that this extinction decreased methane production by about 9.6 million tons per year. This suggests that the absence of megafaunal methane emissions may have contributed to the abrupt climatic cooling at the onset of the Younger Dryas. The decrease in atmospheric methane that occurred at that time, as recorded in ice cores, was 2-4 times more rapid than any other decrease in the last half million years, suggesting that an unusual mechanism was at work.

Examples:

The following are some notable examples of animals often considered as megafauna (in the sense of the "large animal" definition). This list is not intended to be exhaustive:

- Clade Synapsida.
- Class Mammalia.
- Infraclass Metatheria.
- Order Diprotodontia:

  The red kangaroo (*Macropus rufus*) is the largest living Australian mammal and marsupial at a weight of up to 85 kg (187 lb). However, its extinct relative, the giant short-faced kangaroo *Procoptodon goliah* reached 230 kg (510 lb), while extinct diprotodonts attained the largest size of any marsupial in history, up to an estimated 2,750 kg (6,060 lb). The extinct marsupial lion (*Thylacleo carnifex*), at up to 160 kg (350 lb) was much larger than any extant carnivorous marsupial.

- Infraclass Eutheria.
- Superorder Afrotheria.
- Order Proboscidea:

Elephants are the largest living land animals. They and their relatives arose in Africa, but until recently had a nearly worldwide distribution. The African bush elephant (*Loxodonta africana*) has a shoulder height of up to 4.3 m (14 ft) and weighs up to 10.4 tonnes (11.5 short tons). Among recently extinct proboscideans, mammoths (*Mammuthus*) were close relatives of elephants, while mastodons (*Mammut*) were much more distantly related. The steppe mammoth (*M. trogontherii*) is estimated to have commonly weighed around 10 tonnes, making it possibly the largest proboscid, which would make it the second largest land mammal after indricotherines.

- Order Sirenia:

The largest sirenian at up to 1,500 kg (3,300 lb) is the West Indian manatee (*Trichechus manatus*). Steller's sea cow (*Hydrodamalis gigas*) was probably around five times as massive, but was exterminated by humans within 27 years of its discovery off the remote Commander Islands in 1741. In prehistoric times this sea cow also lived along the coasts of northeastern Asia and northwestern North America; it was apparently eliminated from these more accessible locations by aboriginal hunters.

- Superorder Xenarthra.
- Order Cingulata:

The glyptodonts were a group of large, heavily armored ankylosaur-like xenarthrans related to living armadillos. They originated in South America, invaded North America during the Great American Interchange, and went extinct at the end of the Pleistocene epoch.

- Order Pilosa:

Ground sloths were another group of slow, terrestrial xenarthrans, related to modern tree sloths. They had a similar history, although they reached North America earlier, and spread farther north (e.g., *Megalonyx*). The largest genera, *Megatherium* and *Eremotherium*, reached sizes comparable to elephants.

- Superorder Euarchontoglires.
- Order Primates:

  The largest living primate, at up to 266 kg (586 lb), is the gorilla (*Gorilla beringei* and *Gorilla gorilla*, with three of four subspecies being critically endangered). The extinct Malagasy sloth lemur *Archaeoindris* reached a similar size, while the extinct *Gigantopithecus blacki* of Southeast Asia is believed to have been at least two times larger. Some populations of archaic *Homo* were significantly larger on average than recent *Homo sapiens*; for example, *Homo heidelbergensis* in southern Africa may have commonly reached 7 feet (2.1 m) in height, while Neanderthals were about 30% more massive.

- Order Rodentia:

  The extant capybara (*Hydrochoerus hydrochaeris*) of South America, the largest living rodent, weighs up to 65 kg (143 lb). Several recently extinct North American forms were larger: the capybara *Neochoerus pinckneyi* (another neotropic migrant) was about 40% heavier; the giant beaver (*Castoroides ohioensis*) was similar. The extinct blunt-toothed giant hutia (*Amblyrhiza inundata*) of several Caribbean islands may have been larger still. However, several million years ago South America harbored much more massive rodents. *Phoberomys pattersoni*, known from a nearly full skeleton, probably reached 700 kg (1,500 lb). Fragmentary remains suggest that *Josephoartigasia monesi* grew to upwards of 1,000 kg (2,200 lb).

- Superorder Laurasiatheria.
- Order Carnivora:

  The largest extant cats are in genus *Panthera*, including the tiger (*P. tigris*) and lion (*P. leo*). The Siberian tiger (*P. t. altaica*) should be the biggest wild cat according to Bergmann's rule, and has been regarded as such by some but this is disputable. Historically, wild Siberian tigers have declined in size, and they are now smaller than Bengal tigers (*P. t. tigris*); however, Siberian tigers do still tend to be the largest of tigers in captivity. *Panthera* species are distinguished by morphological features which enable them to roar. Larger extinct cats include the American lion (*P. atrox*) and the South American saber-toothed cat (*Smilodon populator*).

  Bears are large carnivorans of the caniform suborder. The largest living forms are the polar bear (*Ursus maritimus*), with a body weight of up to 680 kg (1,500 lb), and the similarly sized Kodiak bear (*Ursus arctos middendorffi*), consistent with Bergmann's rule. *Arctotherium augustans*, an extinct short-faced bear from South America, was the largest predatory land mammal ever with an estimated average weight of 1,600 kg (3,500 lb).

Seals, sea lions, and walruses are amphibious marine carnivorans that evolved from bearlike ancestors. The southern elephant seal (*Mirounga leonina*) of Antarctic and subantarctic waters is the largest carnivoran of all time, with bull males reaching a maximum length of 6–7 m (20–23 ft) and maximum weight of 5,000 kg (11,000 lb).

- Order Perissodactyla:

Tapirs are browsing animals, with a short prehensile snout and pig-like form that appears to have changed little in 20 million years. They inhabit tropical forests of Southeast Asia and South and Central America, and include the largest surviving land animals of the latter two regions. There are four species.

Rhinoceroses are odd-toed ungulates with horns made of keratin, the same type of protein composing hair. They are among the second-largest living land mammals at 850-3,800 kg. Three of five extant species are critically endangered. Their extinct central Asian relatives the indricotherines were the largest terrestrial mammals of all time.

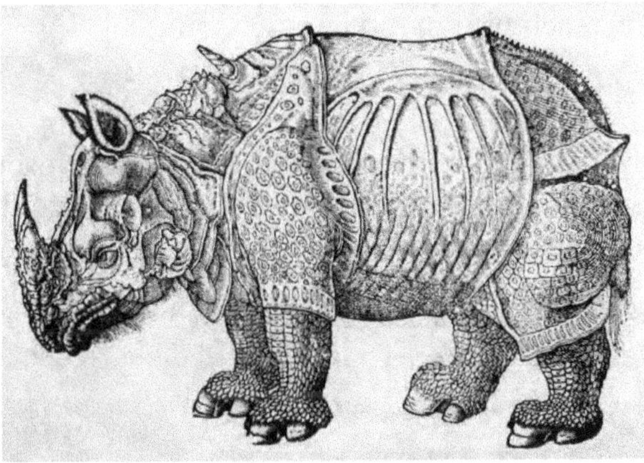

Rhinoceros, from Dürer's woodcut.

- Order Artiodactyla:

Giraffes (*Giraffa spp.*) are the tallest living land animals, reaching heights of up to nearly 6 m (20 ft). The average weight is 1,192 kg (2,628 lb) for an adult male and 828 kg (1,825 lb) for an adult female with maximum weights of 1,930 kg (4,250 lb) and 1,180 kg (2,600 lb) recorded for males and females, respectively.

Bovine ungulates include the largest surviving land animals of Europe and North America. The water buffalo (*Bubalis arnee*), bison (*Bison bison* and *B. bonasus*), and gaur (*Bos gaurus*) can all grow to weights of over 1,000 kg (2,200 lb).

The semiaquatic hippopotamus (*Hippopotamus amphibius*) is the heaviest living even-toed ungulate, mean adult weight is around 1,500 kg (3,300 lb) and

1,300 kg (2,900 lb) for males and females respectively. It and the critically endangered pygmy hippo (*Choeropsis liberiensis*) are believed to be the closest extant relatives of cetaceans. Hippos are among the megafaunal species most dangerous to humans.

- Infraorder Cetacea:

  Whales, dolphins, and porpoises are marine mammals. The blue whale (*Balaenoptera musculus*) is the largest baleen whale and the largest animal that has ever lived, at 30 metres (98 feet) in length and 170 tonnes (190 short tons) or more in weight. The sperm whale (*Physeter macrocephalus*) is the largest toothed whale and one of the largest predators in vertebrate history, as well as the planet's loudest and brainiest animal (with a brain about five times as massive as a human's). The killer whale (*Orcinus orca*) is the largest dolphin.

- Order Pelycosauria (traditional; paraphyletic):

  Cotylorhynchus was a large, big-clawed, herbivorous caseid of Early Permian North America, reaching 6 m (20 ft) and 2 tonnes.

- Order Therapsida:

  Anteosaurus was a headbutting, semiaquatic, carnivorous dinocephalian of Middle Permian South Africa. It reached 5–6 m (16–20 ft) long, and weighed about 500–600 kg (1,100–1,300 lb).

  Lisowicia was an elephant-sized (9 tonne) herbivorous kannemeyeriiform dicynodont of Late Triassic Europe.

- Clade Sauropsida.

- Class Aves.

- Order Struthioniformes:

  The ratites are an ancient and diverse group of flightless birds that are found on fragments of the former supercontinent Gondwana. The largest living bird, the ostrich (*Struthio camelus*) was surpassed by the extinct *Vorombe* of Madagascar, the heaviest of the group at up to (860 kg (1,900 lb)), and the extinct giant moa (*Dinornis*) of New Zealand, the tallest, growing to heights of 3.4 m (11 ft). The latter two are examples of island gigantism.

- Order Gastornithiformes:

  Extinct dromornithids of Australia such as *Dromornis* approached the largest ratites in size. (Due to its small size for a continent and its isolation, Australia is sometimes viewed as the world's largest island; thus, these species could also be considered insular giants.)

- Order Cathartiformes:

  The extinct condor-like teratorn *Argentavis* of South America had an estimated wing span of 5 to 6 m (16 to 20 ft) and a mass of approximately 70 kg (150 lb), making it the best example of a megafaunal flying bird.

- Class Reptilia (traditional; paraphyletic).
- Order Saurischia:

  Saurischian dinosaurs of the Jurassic and Cretaceous include sauropods, the longest (at up to 40 m or 130 ft) and most massive terrestrial animals known (*Argentinosaurus* reached 80–100 metric tonnes, or 90–110 tons), as well as theropods, the largest terrestrial carnivores (*Spinosaurus* grew to 7–9 tonnes; the more famous *Tyrannosaurus*, to 6.8 tonnes).

- Order Pterosauria:

  The largest azhdarchid pterosaurs, such as *Hatzegopteryx* and *Quetzalcoatlus*, attained wingspans around 11–12 m (36–39 ft) and weights probably in the 70–250 kg (150–550 lb) range. The former is thought to have been the apex predator of its island ecosystem.

- Order Crocodilia:

  Alligators and crocodiles are large semiaquatic reptiles, the largest of which, the saltwater crocodile (*Crocodylus porosus*), can grow to a weight of 1,360 kg (3,000 lb). Crocodilians' distant ancestors and their kin, the pseudosuchians (traditional crurotarsans), dominated the world in the late Triassic, until the Triassic–Jurassic extinction event allowed dinosaurs to overtake them. They remained diverse during the later Mesozoic, when crocodyliforms such as *Deinosuchus* and *Sarcosuchus* reached lengths of 12 m. Similarly large crocodilians, such as *Mourasuchus* and *Purussaurus*, were present as recently as the Miocene in South America.

- Order Squamata:

  While the largest extant lizard, the Komodo dragon (*Varanus komodoensis*), another island giant, can reach 3 m (10 ft) in length, its extinct Australian relative *Megalania* may have reached more than twice that size. These monitor lizards' marine relatives, the mosasaurs, were apex predators in late Cretaceous seas.

  The heaviest extant snake is considered to be the green anaconda (*Eunectes murinus*), while the reticulated python (*Python reticulatus*), at up to 8.7 m or more, is considered the longest. An extinct Australian Pliocene species of *Liasis*, the Bluff Downs giant python, reached 10 m, while the Paleocene *Titanoboa* of South America reached lengths of 12–15 m and an estimated weight of about 1,135 kilograms (2,500 pounds).

- Order Testudines:

  The largest turtle is the critically endangered marine leatherback turtle (*Dermochelys coriacea*), weighing up to 900 kg (2,000 lb). It is distinguished from other sea turtles by its lack of a bony shell. The most massive terrestrial chelonians are the giant tortoises of the Galápagos Islands (*Chelonoidis nigra*) and Aldabra Atoll (*Aldabrachelys gigantea*), at up to 300 kg (660 lb). These tortoises are the biggest survivors of an assortment of giant tortoise species that were widely present on continental landmasses and additional islands during the Pleistocene.

- Class Amphibia (in the wide, probably paraphyletic, sense).
- Order Temnospondyli (relationship to extant amphibians is unclear):

  The Permian temnospondyl *Prionosuchus*, the largest amphibian known, reached 9 m in length and was an aquatic predator resembling a crocodilian. After the appearance of real crocodilians, temnospondyls such as *Koolasuchus* (5 m long) had retreated to the Antarctic region by the Cretaceous, before going extinct.

- Class Actinopterygii.
- Order Tetraodontiformes:

  The largest extant bony fish is the ocean sunfish (*Mola mola*), whose average adult weight is 1,000 kg (2,200 lb). While phylogenetically a "bony fish", its skeleton is primarily cartilage (which is lighter than bone). It has a disk-shaped body, and propels itself with its long, thin dorsal and anal fins; it feeds primarily on jellyfish. In these three respects (as well as in its size and diving habits), it resembles a leatherback turtle.

- Order Lampriformes:

  The giant oarfish (*Regalecus glesne*) is the longest bony fish, reaching 11 m (36 ft).

- Order Acipenseriformes:

  The critically endangered beluga (European sturgeon, *Huso huso*) at up to 1,476 kg (3,254 lb) is the largest sturgeon (which are also mostly cartilaginous) and is considered the largest anadromous fish.

- Order Siluriformes:

  The critically endangered Mekong giant catfish (*Pangasianodon gigas*), at up to 293 kg (646 lb), is often viewed as the largest freshwater fish.

- Class Chondrichthyes.
- Order Lamniformes:

  The largest living predatory fish, the great white shark (*Carcharodon carcharias*), reaches weights up to 2,240 kg (4,940 lb). Its extinct relative *C. megalodon* (the disputed genus being either *Carcharodon* or *Carcharocles*) was more than an order of magnitude larger, and is the largest predatory shark or fish of all time (and one of the largest predators in vertebrate history); it preyed on whales and other marine mammals.

- Order Orectolobiformes:

  The largest extant shark, cartilaginous fish, and fish overall is the whale shark (*Rhincodon typus*), which reaches weights in excess of 21.5 tonnes (47,000 pounds). Like baleen whales, it is a filter feeder and primarily consumes plankton.

- Order Rajiformes:

  The manta ray (*Manta birostris*) is another filter feeder and the largest ray, growing to up to 2300 kg.

- Class Placodermi.
- Order Arthrodira:

  The largest armored fish, *Dunkleosteus*, arose during the late Devonian. At up to 10 metres (33 ft) in length and 3.6 tonnes (4.0 short tons) in mass, it was a hypercarnivorous apex predator that employed suction feeding. Its contemporary, *Titanichthys*, apparently an early filter feeder, rivaled it in size. The arthrodires were eliminated by the environmental upheavals of the Late Devonian extinction, after existing for only about 50 million years.

- Class Cephalopoda.
- Order Ammonitida:

  The Late Cretaceous ammonite *Parapuzosia seppenradensis* reached a shell diameter of over 2 m.

- Order Teuthida:

  A number of deep ocean creatures exhibit abyssal gigantism. These include the giant squid (*Architeuthis*) and colossal squid (*Mesonychoteuthis hamiltoni*);

both (although rarely seen) are believed to attain lengths of 12 m (39 ft) or more. The latter is the world's largest invertebrate, and has the largest eyes of any animal. Both are preyed upon by sperm whales.

- Stem-group Arthropoda.
- Order Radiodonta:

  Anomalocarids were a group of very early legless marine arthropods that included the largest predators of the Cambrian, such as *Anomalocaris*. By the early Ordovician they had evolved into giant (for the time) filter feeders, apparently in response to the proliferation of plankton during the Great Ordovician Biodiversification Event. *Aegirocassis* grew to over 2 m in length.

- Order Eurypterida:

  Eurypterids (sea scorpions) were a diverse group of aquatic and possibly amphibious predators that included the most massive arthropods to have existed. They survived over 200 million years, but finally died out in the Permian–Triassic extinction event along with trilobites and most other forms of life present at the time, including most of the dominant terrestrial therapsids. The Early Devonian *Jaekelopterus* reached an estimated length of 2.5 m (8.2 ft), not including its raptorial chelicerae, and is thought to have been a freshwater species.

Meiobenthos

Meiobenthos, also called meiofauna, are small benthic invertebrates that live in both marine and fresh water environments. The term meiofauna loosely defines a group of organisms by their size, larger than microfauna but smaller than macrofauna, rather than a taxonomic grouping. In practice, that is organisms that can pass through a 1 mm mesh but will be retained by a 45 μm mesh, but the exact dimensions will vary from researcher to researcher. Whether an organism will pass through a 1 mm mesh will also depend upon whether it is alive or dead at the time of sorting.

The term meiobenthos was first coined in 1942 by Mare, but organisms that would fit into the meiofauna category have been studied since the 18th century. A comprehensive text on meiofauna is introduction to the study of meiofauna by Higgins and Thiel.

Collecting the Meiobenthos

Meiofauna are most commonly encountered in sedimentary environments in both marine and fresh water environments, from the littoral to the deep-sea. They can also be found on hard substrates living on algae, the phytal environment, and sessile animals (barnacles, mussel beds, etc.).

Sampling Methodologies

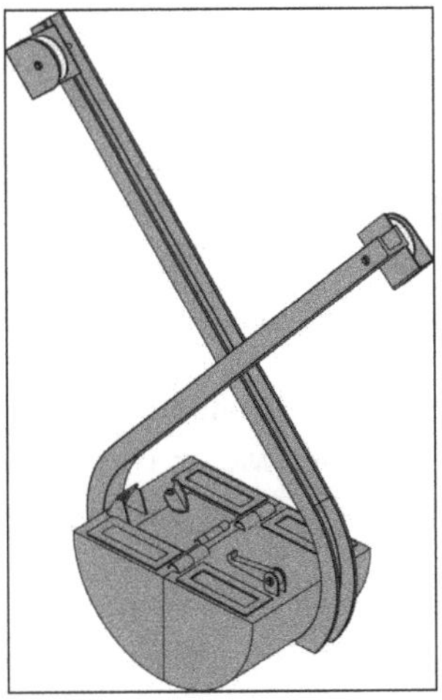

Van Veen grab.

Sampling the meiobenthos is clearly dependent upon the environment and whether quantitative or qualititative samples are required. In the sedimentary environment the methodology used also depends on the physical morphology of the sediment. For qualititative sampling within the littoral zone, for both coarse and fine sediment, a bucket and spade will work. In the sub-littoral and deep water some form of grab (like the Van Veen Grab Sampler) is required, although a fine mesh (about 0.25 mm or less) would work also.

For the quantitative sampling of sedimentary environments at all depths a wide variety of samplers have been devised. The simplest is a plastic syringe with the end cut off to form a piston corer which can be deployed in the littoral zone, or in the sub-littoral using SCUBA gear. Generally the deeper the water the more complicated the sampling process becomes. For sampling the meiofauna on hard substrates, phytal and epizooic environments, the only practical methodology is to cut or scrape off a known area of substrate and place it in a plastic bag.

Extraction Methodologies

There are a wide variety of methods for extracting meiofauna from the samples of their habitat depending upon whether live or fixed specimens are required. For extracting live meiofauna one has to contend with the large number of species that cling or attach themselves to the substrate when disturbed. In order to get the meiofauna to release their grip there are three methodologies available.

The first, and simplest, is osmotic shock, this is achieved by submerging the sample in fresh water (clearly this will only work for marine samples) for a few seconds. This will cause the organisms to release after which they can then be shaken free from the substrate and filtered out through a 45 μm mesh and immediately returned to fresh filtered seawater. Many organisms will come through this process unharmed as long as the osmotic shock does not last too long.

The second methodology is the use of an anaesthetic. The preferred solution for meiobenthologists is isotonic magnesium chloride (7.5g $MgCl_2 \cdot 6H_2O$ in 100 ml of distilled water). The sample is immersed in the isotonic solution and left for a period of 15 min, after which the meiofauna are shaken free of the substrate and again filtered out through a 45 µm mesh and immediately returned to fresh filtered seawater.

The third methodology is Uhlig's seawater ice technique. This relies on the organisms moving ahead of a front of ice cold seawater moving down through the sample ultimately forcing them out of the sediment. It is most effective on samples from temperate and tropical regions.

For major studies where large numbers of samples are collected concurrently, samples are normally fixed using 10% formalin solution and the meiofauna extracted at a later date. There are two main extraction methodologies. The first, decantation, works best with coarse sediments. Samples are shaken in an excess of water, the sediment is briefly allowed to settle and the meiofauna filtered off. The second methodology, the flotation technique, works best with finer sediments where the mass of the sediment particles is close to that of the meiofauna. The best solution to use is the colloidal silica, Ludox. The sample is stirred into the Ludox solution and left to settle for 40 min, after which the meiofauna are filtered out. In fine sediments, extraction efficiency is improved with centrifugation. With both methodologies repeated extractions should be made (at least three) with each sample to ensure that at least 95% of the fauna is extracted.

Microfauna

Microfauna refers to small, often microscopic animals, especially those inhabiting the soil, an organ, or other localized habitat. Single-celled protozoans, small nematodes, small unsegmented worms, and tardigrades (eight-legged arthropods) are the most common components of microfauna. Many inhabit water films or pore spaces in leaf litter and in the soil, feeding on smaller microorganisms that decompose organic material.

IMPORTANCE OF FLORA AND FAUNA

Maintains Ecological Balance

Without flora and fauna, humans cannot exist. The flora generates and releases oxygen, which is needed by the fauna for respiratory purposes. In return, the fauna produces and releases carbon dioxide, which is needed by the flora for photosynthesis. It's a symbiotic kind of relationship. In the same line, humans cannot get by without both flora and fauna. The oxygen that we breathe in comes from the flora, and the carbon dioxide we exhale is vital for the flora.

Also, humans benefit a lot from flora and fauna in regards to sources of food, medicine, and water. Our main source of food emanates from plant and animal species. Over 90% of medicine we use to cure diseases comes from flora. If it were not for the flora, there would be no water, which means we would not exist right now.

Animals also maintain the equilibrium across the board by predating on plants and other animals that could have otherwise exploded in regards to population. They also enable other rare species of plants to mushroom by pollinating other plants. Animal droppings is a source of fertilizer for plants. When animals die, they act as supplemental mineral for plants. Also, microorganisms on the Earth' surface benefit from abundance of food as a result of animal droppings.

Aesthetic Value

It's a no-brainer that humans love and appreciate nature. Many like spending time in outdoor settings such as forests, natural areas, parks and other green spaces because of their aesthetic value. This aesthetic value is mainly contributed by spread of flora and fauna. Statistics tell the story, each year, up to half a billion people visit beautiful protected sceneries like national parks, recreation areas, indigenous forests, historic sites, wildlife refuges and wild and scenic rivers to experience the beauty of the landmarks. This further underlines the significance of flora and fauna to our day to day lives.

Expands Local Economies

Flora and fauna contribute mightily to most world economies regarding tourism. For example, flora and fauna of Amazon attract numerous scientists and explorers. It's estimated that the Amazon rainforest contributes about $50 million dollars to the Brazil economy. Exotic vacation locations like the Caribbean, Bahamas, Panama, Indonesia and so on attract more tourists than any country due to the widespread flora and fauna.

ENDANGERED FLORA AND FAUNA

The ecosystem is a consistent, harmonized holistic organisation functioning according to its own laws. It sustains on its own laws and maintains its own balance independent of outsiders. Human intervention has posed a danger to the ecosystem and endangered many species of plant and animal life.

The pernicious destruction of the rainforests has destroyed half of the world's species of plants, animals and microorganisms. Birds and animals are in part responsible for keeping the rainforest alive. They spread seeds through their faeces and allow new trees to germinate. Rapid destruction of trees has affected bird habitat and correspondingly many plant species.

The indiscriminate killing of sharks for shark meat has created a dangerous overpopulation of sting rays. Similarly, unlicensed whaling in Japan has caused many species of whales to be seriously endangered.

The Great Barrier Reef, home to millions of the ocean's species, is being threatened as nitrate-laden water is dumped into the Great Barrier Reef. Overfishing of the Giant Triton and sharks, has caused disruption to food chains vital to life on the reef. Shipping accidents are a threat, to the reef as are foreign species discharged from ships creating a biological hazard to the Great Barrier Reef.

There are lots of factors that impact extinction of world's fauna and flora which include the human quest for money and power. This has created a world in which the ecological balance of flora and fauna is being gradually destroyed. In extreme instances, this has caused the extinction of entire species due to destruction of habitat and food sources.

Some of the factors that can be identified for this calamity are excessive use of insecticides which are absorbed by the ground water. This infiltrates into the streams and waterbodies which when consumed by animals can be cause death.

There are hundreds of species of animals throughout the world which are fast disappearing because of human interference in their natural habitat. The more animals and plants we lose, the fewer there are to contribute to individual ecosystems. The Spanish lynx, the North American buffalo, the Siberian Tiger, the Royal Bengal Tiger, the Kola Bear, are just a few of the animals we are fast losing. The more individual ecosystems are damaged, the greater the negative effects to the Earth's environment as a whole.

References

- What-are-flora-and-fauna: conserve-energy-future.com, Retrieved 23 February, 2019
- Cullen, james; knees, sabina; cubey, h suzanne, eds. (2011), the european garden flora: a manual for the identification of plants cultivated in europe, both out-of-doors and under glass. Vols 1–5, cambridge, uk: cambridge university press, isbn 9780521761673 (5 volume set)
- What-are-flora-and-fauna: conserve-energy-future.com, Retrieved 10 July, 2019
- Sorkin, b. (2008-04-10). "a biomechanical constraint on body mass in terrestrial mammalian predators". Lethaia. 41 (4): 333–347. Doi:10.1111/j.1502-3931.2007.00091.x
- Importance-flora-and-fauna, ecosystem: eartheclipse.com, Retrieved 31 May, 2019
- Giere, olav (2009). Meiobenthology. The microscopic motile fauna of aquatic sediments, 2nd edition, springer. Isbn 978-3-540-68657-6
- Microfauna, science: britannica.com, Retrieved 13 March, 2019

Permissions

All chapters in this book are published with permission under the Creative Commons Attribution Share Alike License or equivalent. Every chapter published in this book has been scrutinized by our experts. Their significance has been extensively debated. The topics covered herein carry significant information for a comprehensive understanding. They may even be implemented as practical applications or may be referred to as a beginning point for further studies.

We would like to thank the editorial team for lending their expertise to make the book truly unique. They have played a crucial role in the development of this book. Without their invaluable contributions this book wouldn't have been possible. They have made vital efforts to compile up to date information on the varied aspects of this subject to make this book a valuable addition to the collection of many professionals and students.

This book was conceptualized with the vision of imparting up-to-date and integrated information in this field. To ensure the same, a matchless editorial board was set up. Every individual on the board went through rigorous rounds of assessment to prove their worth. After which they invested a large part of their time researching and compiling the most relevant data for our readers.

The editorial board has been involved in producing this book since its inception. They have spent rigorous hours researching and exploring the diverse topics which have resulted in the successful publishing of this book. They have passed on their knowledge of decades through this book. To expedite this challenging task, the publisher supported the team at every step. A small team of assistant editors was also appointed to further simplify the editing procedure and attain best results for the readers.

Apart from the editorial board, the designing team has also invested a significant amount of their time in understanding the subject and creating the most relevant covers. They scrutinized every image to scout for the most suitable representation of the subject and create an appropriate cover for the book.

The publishing team has been an ardent support to the editorial, designing and production team. Their endless efforts to recruit the best for this project, has resulted in the accomplishment of this book. They are a veteran in the field of academics and their pool of knowledge is as vast as their experience in printing. Their expertise and guidance has proved useful at every step. Their uncompromising quality standards have made this book an exceptional effort. Their encouragement from time to time has been an inspiration for everyone.

The publisher and the editorial board hope that this book will prove to be a valuable piece of knowledge for students, practitioners and scholars across the globe.

Index

www.ingramcontent.com/pod-product-compliance
Lightning Source LLC
LaVergne TN
LVHW080605210926
848485LV00003B/109

9781641726696